The
Yorkshire
Countryside

A Landscape History

The
Yorkshire
Countryside

A Landscape History

Richard Muir

KEELE UNIVERSITY PRESS

To the memory of Jack Ravensdale,
a gifted historian and a fine gentleman

© Richard Muir, 1997

KEELE UNIVERSITY PRESS
22 George Square, Edinburgh

Typeset in Palatino Light
by Pioneer Associates, Perthshire.
printed and bound in Great Britain

A CIP record for this book is available
from the British Library

ISBN 1 85331 198 7

The right of Richard Muir to be identified as
The author of this work has been asserted in
accordance with the Copyrights, Designs and
Patent Act (1988)

All photographs by Richard Muir
unless otherwise stated

Contents

Introduction

THERE ARE SEVERAL APPROACHES to landscape history; some concern subtle differences of emphasis – like the difference between Carl Sauer's cultural approach to the differentiation of the land surface and Oliver Rackham's treatment of landscape in terms of historical ecology. There are others which regard the landscape from quite different standpoints and consider it in terms of the ways in which it is perceived by different sorts of people at different times, or the way in which it expresses the social patterns of power and influence, as well as still other approaches which see the landscape as a metaphor for social conditions. In addition, there are the perspectives, developed by geographers, which engage with the aesthetic aspects of landscape, as pioneered in Britain by Jay Appleton and elaborated by Denis Cosgrove.

Bearing in mind the psychological associations of place and territory, it can be said that Yorkshire is not only a land, it is also a state of mind – one which laps freely across the boundaries of modern administration to be shared by Yorkshire people reassigned to ill-fated Humberside and to Cleveland, and one which lives on in the emotions and self-images of countless expatriates. Yorkshire is both not a county and much more than a collection of counties. The word itself invokes powerful imagery. Outsiders who have but the vaguest impressions of, say, Leicestershire or the Chilterns are sure to have a distinct mental image of Yorkshire. It might be centred on visions of wind-lashed moorland, smoking chimneys, or tough, blunt people; it could be superficial and misinformed; but everyone has an image of Yorkshire – probably a firm opinion, too. The Yorkshire self-image seems to have been established for quite some time; it was in 1875 that James Burnley remarked that: 'They were Yorkshire to begin with and Yorkshire they will remain to the close. Wave after wave of change may pass over them; but they will stand firm and immutable in their adherence to the traditions and customs of their forefathers.' Burnley considered that the Yorkshireman was the most 'birth-proud' member of the human race, while Dellheim felt that 'there was no doubt about one of his [or her?] striking characteristic – regional pride'.

This quality of Yorkshireness was far from being indivisible. In terms of both people and place, Barnsley, Swaledale and Holderness are hugely different – and these differences were intensified by the effects on local cultures and landscapes of the Industrial Revolution. These contrasts too were neatly encapsulated by Dellheim: 'Writers often coupled together Leeds and York to symbolize the contrast between Yorkshire past and present. York was "hoary with memories of past generations and quiet with the restfulness that belongs to age"; Leeds was, in a revealing phrase, "grimy with prosperous toil, darkening the sky with the smoke of its factories".' The two great centres are still different, but in different ways, with York making a living from its past, and Leeds – once more an exciting place – transformed with post-modernist buildings and pioneering a pathway into the post-industrial age. When we consider peoples' perception of Yorkshire and its landscapes, an extra dimension of difference is superimposed upon those which exist in the 'real' world. The perceptions tend to be as schizophrenic as the real landscapes, a juxtaposition of towering chimneys and smoke-wreathed valleys with bracing moorlands and idyllic dales. Stereotypes were dispersed and reinforced by the rise of the regional novel. Pocock explains that: 'The environment now considered characteristic of the North had become increasingly apparent in the second quarter of the last century as industrialization, aided by the railway, changed the basis of the country's wealth from agriculture to manufacturing, thus bringing increased importance to the North'. The image of the North as a place of smoke, disfigured scenery and mean townscapes, peopled by folk with temperaments which oscillated around the extremes, which was disseminated by the early nineteenth-century novelists continued to be broadcast into modern times, so that Pocock found that: 'The search has revealed a consistency in projection which may be recognised as a key contributor to the total image-geography of the North, which in many ways is now false or alien geography.'

Yorkshire should amply reward studies of landscape related to the perception of place, culture and self. No less rewarding would be studies relating to the landscape as an expression of the unequal relationships of power within society. Explorations of this kind could be conducted at any cross-section in time. They might, for example, concern the feudal countryside and the division of territory between the royal forest, the deer parks or the aristocracy, the areas of manorial demesne and the communal strip fields, meadows and commons. Alternatively, they could consider episodes in time and place like the inundation of the former village of Hinderskelfe in the creation of an ornamental lake at Castle Howard in the years around 1700, the 'improvement' of farmland across a great tract of the wolds by Sir Christopher Sykes of Sledmere a century later, or the roles of the Dukes of Norfolk in financing the development of coal mining on their estates near Sheffield in the decades around 1800. All such studies would underline the points made by Williamson and Bellamy that:

> As people went about their business, providing themselves with food and shelter, they inevitably left an impression on their surroundings. But the nature of this impression was determined by more complex factors than the simple need to satisfy material requirements.

The way in which wealth was distributed in society, the relationship between the various classes, and the changing demands of the national economy all influenced the appearance of the countryside. The present landscape is the product of the activities of innumerable different communities over thousands of years. Its varied fabric and features pose a host of questions about the nature of societies of the past.

Equally, one could pursue an analysis of the Yorkshire landscape which was guided by eco-logical principles. It might explore notions of the carrying capacity of the Mesolithic landscape in terms of game animals and their human predators, and consider the shifting equilibrium between population and available farmland – with the filling and then overfilling of the countryside apparently causing episodes of retreat and aban-donment even in Neolithic times. Any illusions that human progress could be represented by an ever-steepening upward curve have been dispelled by a wealth of archaeological and documentary work which reveals phases of population growth, expansion and land hunger being succeeded by crisis and retreat. Studies of the consequences of climatic change and the human abuse of the environment in, say, middle Bronze Age, late Roman or fourteenth-century times would surely contain messages for modern society in Yorkshire.

While reference will be made to each of these perspectives, a broad approach has been adopted here, with the aims of charting the evolution of settlements and countrysides in Yorkshire as far as the earlier phases of the Industrial Revolution; of reconstructing the countrysides as they appeared and evolved during successive periods; and of considering particular facets of the scene, such as dwellings and strongholds. The text closes with the early stages of the Industrial Revolution because it is considered that, with the acceleration of change associated with industrialisation, the pattern of development assumed a degree of complexity which renders attempts to analyse and interpret events, patterns and processes in just two or three chapters impossible. No attempt would be made to suggest that 'the modern' has no place in a study of landscape history, but that issues of accelerated urbanisation, Fordism, time/space compression and several other crucial aspects cannot be treated adequately in the space of a few pages. The topic of the industrialisation of Yorkshire would require a book of at least the length of this one. The use of a computer and scanner have enabled me to use an earlier text, *Old Yorkshire*, as a framework on which to hang some ideas; in a great many places the framework is abandoned and rebuilt and this is certainly not a revision or expansion of that text. I am indebted to two colleagues, pollen analyst and biogeographer Margaret Atherden, and geomorphologist Catherine Fiske, for their helpful comments on the first two chapters, where any faults still remaining are entirely my own responsibility. In writing this book, I have been very conscious of my debts to authors on the Yorkshire landscape whom I have never met or know only slightly, and, particularly, to friends and fellow landscape historians who have helped to shape my outlook: Christopher Taylor, Tom Williamson, Liz Bellamy and the late Jack Ravensdale; this work is dedicated to Jack's memory.

REFERENCES

J. Appleton, *The Experience of Landscape* (London: 1975).

J. Burnley, *West Riding Sketches* (1875), quoted in C. Dellheim, below.

D. Cosgrove and S. Daniels, *The Iconography of Landscape* (Cambridge: 1988).

C. Dellheim, 'Imagining England: Victorian views of the North', *Northern History,* 23 (1987), pp. 216–230.

D. C. D. Pocock, 'The novelist's image of the North', *Transactions of the Institute of British Geographers,* NS 4 (1979) pp 62–76.

O. Rackham, *The History of the Countryside* (London: 1986).

C. O. Sauer, 'The morphology of landscape', *University of California Publications in Geography,* 2 (1925), pp. 19–54.

T. Williamson and E. Bellamy, *Property and Landscape* (London: 1987).

CHAPTER ONE

The Physical Environment

IT IS SOMETHING OF A CONVENTION for accounts of landscape history to begin with a discussion of the physical environment. This is not because the physical setting determines the direction of human development, but that it tends to bracket the activity range within which endeavours may be expected to succeed. The position of these brackets is not fixed, for they will expand, shift or contract through time, while development will also be conditioned by factors such as culture, perception and the technological level achieved by the society concerned. The physical environment should not be regarded as a constant, with an existence that is independent of human activities. Firstly, during the current phase of uninterrupted human occupation of Yorkshire, climatic conditions have ranged from the subarctic to the temperate; on at least two occasions the entire direction of socio-economic development was changed by marked deteriorations of climate. Secondly, the human relationship with the physical context is interactive; on several occasions the ability of the environment to sustain human settlement has been undermined by ecologically detrimental exploitation, with, for example, a severe episode of soil erosion caused by arable farming apparently occurring towards the end of the Roman period in the Yorkshire Wolds.

Yorkshire is not a region of the classical geographical kind and, in this text, the term 'region' is only applied to Yorkshire for the sake of convenience. A case of kinds can be made for regarding Yorkshire as a microcosm of England as a whole, while there is a stronger case for stressing the regional characteristics of its main components: the Dales, the North York Moors, the Yorkshire Wolds and the Vales of York and Pickering. Yorkshire epitomises the cultural distinction between upland and lowland Britain which has been fundamental to an appreciation of British landscape history since the concept was first articulated by Sir Cyril Fox in 1932. In terms of its upland scenery, Yorkshire is something of an English microcosm, with the Carboniferous rocks of the Pennine Dales, the Jurassic geology of the North York Moors, and the Cretaceous structures of the Wolds creating a trio of markedly distinctive regions. Variation between the vales in the lowlands is less marked, though proximity to the North Sea enhances the identity of Holderness.

The landscape historian's interest in the physical setting is not focused solely on the role of this setting as a context for human development. The quality of evidence available for analysis is conditioned by geological and topographical constraints, with, for example, the Bunter (Sherwood) sandstone in the south of Yorkshire producing excellent crop-marks, the Carboniferous limestone being associated with a good revelation of prehistoric field and settlement evidence, while the aerial photographic results from the coarse or silty acidic soils of the Carboniferous sandstone are generally disappointingly barren. In this way, variations in the manner in which different physical contexts reveal evidence of human colonisation could produce distortions in interpretations of the distribution of communities. Also, it is important to remember that factors like erosion and sedimentation are continuous processes which are no less active today than they were a thousand years ago. In parts of Yorkshire fossilised landscapes of colonisation are preserved, to the extent that the uplands of the North York Moors have been likened to an open-air museum of the Bronze Age. However, such areas could still have been marginal or semi-peripheral to the main theatres of action, so that the cairn-fields, barrows and boundary earthworks seen on the moors probably existed on the outer zones of the colonised areas, with the main agricultural and settlement hubs lying in the valley bottoms. There they will now lie buried beneath several metres of alluvium or under debris eroded and transported from the slopes above.

Discussions of a geological nature can pose unfamiliar conceptual problems for the landscape historian; while rocks found in a few places in Yorkshire date from around 450 million years ago, there was no Yorkshire then, just a vast sheet of ocean; 400 million years ago the land was part of a great desert continent; and about 100 million years later it was blanketed by tropical forest and swamp. The relative movements of the continental plates complicate the picture, but can be overlooked in a human-orientated study. The geological structure of Yorkshire is relatively uncomplicated and, with its older, tougher rocks to the west and the younger, softer beds to the east, Yorkshire mirrors the general pattern of England as a whole. In a few places in the Pennines – exposed in the walls of the magnificent gorge at Ingleton, beside Malham Tarn, in the Howgill Fells north of Sedbergh, where ancient green slates are exposed, and in Ribblesdale – one can find some more ancient rocks but, in the main, the rocks of the Pennines are a legacy of the Carboniferous period, some perhaps 360 million years old, some rather younger. The Carboniferous landscapes were themselves the results of previous cycles of orogenesis, uplift, denudation and sedimentation. Long ago, ranges of mountains had been thrown up and masses of molten granite had been injected into the roots of the scenic blocks. Erosion had then stripped away from these mountains rock that was once thousands of metres thick, eventually creating an uneven, sagging plain, traversed by river valleys and punctuated by hills formed from the stumps of worn-down highlands. Then the clear, tropical waters of a shallow sea invaded the plain; as its population of crustaceans, corals, sea lilies and micro-organisms died, their calcareous skeletons fell to the seabed. This steady rain of shells and skeletons continued for millions of years and, in places, the white sea-floor sludge accumulated to

depths of over 150 metres. Compressed by the weight of water and sediments above, the shell debris eventually produced the tough, silvery Carboniferous or Great Scar limestone on which is etched most of the distinctive scenery of the Yorkshire Dales.

Particles of debris eroded from the adjacent land mass were deposited in the tropical sea, which experienced phases of accelerated deposition when the waters became clouded by mud and silt before clearing again. Gradually, the waters became shallower, the sea floor blanketed in a deepening shell debris. Into the fringing tropical lagoons were deposited muds washed down by rivers from the lands lying to the north; fluctuations in the cycles of erosion and deposition resulted in strata of mud, shale, sandstone and limestone accumulating on the seabed in alternating layers, some of them just a couple of metres thick. Interbedded with these were a few narrow seams of coal, each created when swamp forest colonised a shallow lagoon. These intricate beds form the rocks of the Yoredale Beds, sometimes seen standing above the Great Scar limestone in places in Wensleydale or Wharfedale and creating a staircase terrain which forms as differential weathering and erosion reveal contrasts in the resistance of adjacent rocks.

Somewhere to the north of what one might imagine as being 'ancestral Yorkshire', and spanning the area that now lies between Norway and the Hebrides, was a great and rapidly eroding land mass. In the latter part of the Carboniferous period, the rivers flowing into the shrinking sea were charged with masses of coarse sands, deposited in immense thicknesses as deltas and sandbanks around the margins of the sagging sea floor. Compacted and cemented, these sands gave rise to the distinctive Millstone Grit. (In due course, much of the sandstone blanket would be eroded away from the Pennines, surviving as islands forming tough gritstone slabs capping peaks like Ingleborough or Pen-y-Ghent, or covering more extensive areas in Nidderdale and parts of the valleys of the Swale, Ure and Wharfe). The predominant aspects of the physical character of the dales is produced by the contrasting characters of the dry, bare green pastures and the rock exposures in hillside scars and pavements of the

Millstone Grit forms a hard capping on some Pennine summits, like Pen-y-Ghent and Ingleborough. Below are the intricate beds of the Yoredale Series and then the Great Scar limestone.

Great Scar limestone country and the darker, heather moors and moist, tree-edged pastures of the gritstone lands – the marked scenic and ecological differences being attributable, as noted, to changing patterns of sedimentation affecting a tropical sea around 300 million years ago). The period towards the close of the Carboniferous era was marked by the formation of the Coal Measures. These beds reveal a fluctuating cycle of terrestrial and marine conditions, with peat-bed formation on the coastal plains being succeeded by the marine transgressions which would seal the peat under strata of mussel beds and limestones. Next, river-borne deposits of mud might blanket the marine deposits as the sea shallowed, with rivers then depositing sands on the emergent shoreline. Forest would then colonise the swampy ground in the coastal zone, and the cycle of subsidence, deposition and emergence would be renewed. In time, the compression of the beds of plant material produced the coal seams which were destined to permit a major transformation of the West Riding during the steam-powered phase of the Industrial

Revolution, while the intervening sandstones are still quarried to provide several notable building stones.

After the formation of the Coal Measures, there was a major episode of orogenesis; the precursor of the Pennines was uplifted as an imposing mountain chain, while gigantic faults split the chain into vast blocks. Sheets of molten rock were injected through the fractures, with one such sheet of dolerite rock of great thickness forming the Whin Sill, exposed beyond the north of Yorkshire as the geological rampart which carries Hadrian's Wall. Other such injections of molten rock along lines of weakness in the limestone, occurring at various times, produced economically significant mineral veins yielding lead, calcite, barytes and fluorspar. For many millions of years after the orogenesis, the 'ancestral Pennines' existed as an island; eventually, around 70–80 million years ago, the mountains probably subsided completely into the ocean, when a thick blanket of chalk would have been deposited across the submerged landscape. Then, culminating a mere 10–20 million years ago, ripples from the Alpine orogenesis elevated the Pennines again and reopened the old fault-lines which bounded the upland blocks. As the Pennines gradually rose, by perhaps 500–700 metres, so new rivers carved their courses into the slowly rising landscape. If the reborn Pennines had a complete chalk cover – which is debatable – the forebears of the rivers of the dales would soon have stripped it away, though the Cretaceous submergence is still apparent in the formation of the chalk countryside of the Wolds and the white cliffs exposed at places like Bempton and Flamborough Head.

At about the time that the grits, sandstones and coal deposits were accumulating around what is now the area of the Pennines, the area to the east formed part of a vast ancestral North Sea basin; as the basin sagged and sank, so its bed accumulated great thicknesses of sediments. The nature of these materials varied according to relative levels of land and sea, but in the course of the last 225 million years the centre of the basin accumulated sediments to the depth of approximately 5,000 metres. Subsidence has been greatest in this centre, partly explaining why the

rock layers which form the eastern part of Yorkshire tend to dip downwards towards the east. Erosion has stripped away great thicknesses of rock, cutting across the eastward-dipping strata to expose a succession of beds whose rocks become progressively younger as one comes closer to the shores of the North Sea. At the eastern edge of the Dales, the Millstone Grit is succeeded by the younger Magnesian limestone, formed as salts evaporated from lobes of the desert-girt sea. The change is instantly apparent in the landscape as farmsteads of gritstone give way, at Knaresborough, to dwellings displaying the softer texture and creamy tone of the younger limestone. The Magnesian limestone forms a historically valuable, though narrow, western hem to the Vale of York; to the east lies a broader zone of desert sandstones, deposited by a river which originated in 'France' and which, when in spate, had the power to surge across the arid lands carrying a load of silts, sand and pebbles. In general, the geological strata of the Vale of York lie buried by several metres of glacial till. At the eastern margins, one leaves the desert sandstones and crosses a narrow band of shales, ironstones and sandstones which record the next geological chapter, when the land was invaded by a muddy sea. Trunks of trees washed down from the adjacent land mass and sunk into the sea-floor sediments were eventually converted into jet, whose commercial significance rose when the death of Prince Albert inspired a fashion for ornaments and jewellery in black Whitby jet.

The geological origins of the North York Moors can be summarised as follows, Approximately 208–13 million years ago, fine silt and mud sinking to the floor of a sea were deposited to form the Lias which is exposed in the floor of the Esk valley and Rosedale. Trees growing in the marginal swamps were the source of jet, while associated iron-rich deposits provided the basis for the iron-mining industry of Cleveland. Continuing deposition caused the sea to become shallower and the area was overrun by an enormous river delta, the sands and silts deposited in deltaic formations being compressed to form the Jurassic sandstones of the North York Moors. The silting-up of this murky sea brought a brief return to terrestrial conditions; there followed a submergence beneath

The tough rocks of the North York Moors present an almost cliff-like face to the gentle lands of the Vale of York in the Hambledon Hills at Sutton Bank. In the middle distance, the glacial lake of Gormire nestles in a hollow well above the level of the plain.

a coral-rich sea – which produced the coralline grits and limestones of the North York Moors. Then the Kimmeridge clay was deposited, almost all of which was subsequently removed by erosion. During the Alpine orogeny, fracturing of the Jurassic rocks took place, with the injection into the faults of a volcanic rock known in the north as 'whinstone'. In the Yorkshire Wolds, chalk surviving from the subsequent Cretaceous submergence, can be almost 500 metres thick, though elsewhere in Yorkshire, as we have seen, erosion has completely stripped it away. The great earth movements which re-created the Pennines rippled the rocks in the north-east of Yorkshire, with an uplifting of the land giving rise to the uplands of the North York Moors, and a downfold producing the Vale of Pickering. A recent field guide produced by the Yorkshire Geological Society notes that:

In the Tertiary pressure from the east gently folded the chalk into a saucer-like basin, producing scarps in the west and south Wolds and reactivating the older marginal faults along the Howardian–Flamborough Fault System, resulting in a compressional/extensional fracture zone running east–west across the Wolds and exposed on the coast. The chalk was compressed and recrystallized to form the resistant Flamborough headland. Marine erosion has exploited the many minor faults associated with the crush belt, forming the magnificent coastal scenery of arches, stacks, caves and coves.

Less than 2 million years ago, the climate underwent a severe deterioration and Yorkshire experienced episodes of glaciation succeeded by interglacial phases of quite temperate conditions. In practice, it is difficult to identify the effects of particular ice ages, because each great glaciation would tend to mask or obliterate the land-forms created in the preceding glaciation. It appears that in the course of the penultimate ice age, occurring between about 200,000 and 125,000 years ago, ice from Scotland streamed southwards into the Eden valley; ice from the Irish Sea

The main waterfall in Gordale Scar, frozen during the exceptionally harsh winter of 1985–6, evokes images of Ice Age Yorkshire

completely engulfed the Isle of Man, while Scandinavian ice overrode the eastern shores of Yorkshire. This was followed by an interglacial period that was sufficiently warm to allow (on the evidence of the discovery of a tooth) an ox to flourish near Appleby in the Eden valley – and, one can assume, mammoth, horses. woolly rhinoceros and wolves to flourish in Yorkshire. The last great ice age, the Devensian, running from about 80,000 to, say, 12,000 years ago, is the most clearly understood. Glacial erosion was particularly ferocious in the Pennines, and across the north of England there was a great merging of Scottish, Lake District, Cheviot, Scandinavian and Pennine ice. To the north of the region, the great fault-bounded section of the Pennines between the Tyne and Tees headwaters, known as the Alston block, intercepted the snow-laden clouds with its craggy western face. An independent ice-cap developed on the high plateaux, and glaciers radiated outwards from this gathering ground. In the Pennines further south lies another great fault-girt mass, the Askrigg block, and here, around the headwaters of the Ure and the Eden, was another ice-accumulation centre. In parts of the Dales, quite severe glacial erosion took place, producing U-shaped valleys and truncated spurs,

though not the knife-edged ridges or arêtes created in the Lake District. Relative location was crucial in determining the intensity of glacial erosion. King wrote that: 'The ice did not produce much over-deepening, but was probably responsible for the formation of the limestone scars that are so characteristic of the dales, by scouring off the weathered mantle', and she explained that the mass of Buckden Pike:

diverted the ice to north and south to flow respectively down upper Wharfedale and Bishopdale. The ice in the latter valley had a steep gradient at the valley head, and was able to scour the valley bottom to such an extent that the main valley now hangs to its short tributary... Bishopdale thus forms a strong contrast to its southern neighbour Walden, which was protected from the ice by the bulk of Buckden Pike, and which is now a V-shaped valley, showing very little sign of glacial erosion.

One of the beautiful calcite formations visible to the public at the remarkable Stump Cross caverns

While actively eroding in the valleys of the dales, the glaciers of the last glaciation lost momentum as they reached the flatter gradients of the Vale of York, but, merging and extending, they reached locations more than 10 miles (16 km) to the east and south of York. At this time, a local ice-cap completely enveloped the Lake District; in the central and northern Pennines, however, several of the loftier summits – Buckden Pike, Pen-y-Ghent, Ingleborough and a few others – stood above the encircling ice. Meanwhile, in the more southerly Pennine regions, snow blanketed the moors, though ice-caps were absent. The North York Moors were completely overridden by ice during the penultimate glaciation, but during the last glaciation, ice-free but snow-bound, they were skirted by vast 'rivers' of ice which had their sources variously in Scotland, the Pennines, and also in Scandinavia – where the ice had amassed to a thickness of approximately 2,500 metres. To the east, the Scandinavian ice overran the line of the present coast, extending inland to Lealholm in Eskdale and to Wykeham in the Vale of Pickering, while between Flamborough Head and Spurn Head the ice completely covered the old shoreline and advanced inland to the vicinity of Beverley. Debris dumped by the Scandinavian ice prevented the Derwent from flowing eastwards into the North Sea and it developed a new course, flowing southwards and westwards to join the drainage system of the Ouse.

So far as the later human colonisation was concerned, the glaciation was most significant through the redistribution of soil, with the economic inequalities between upland and plain being intensified by the removal of soil from the relatively impoverished upland environments and its deposition in the lowlands. The glaciers descending from the elevated centres of ice accumulation followed the courses of the pre-existing rivers, occasionally gouging deeply into the old river-beds, plucking rocks from the ever-steepening valley sides, capturing the debris falling from the frost-shattered slopes above, and scouring away the accumulations of soil beneath. The limestone pavements of the plateaux of the Pennines, with the blocks of stone or clints being defined by open crevices or grikes, may partly reflect overgrazing and the tendency for soil accumulations to be flushed away through the fissure networks by rainfall, but glaciation was largely responsible for the stripping of upland soils and exposing the pavements to attack by running water. Other factors were involved, and Goudie considered that:

> For good pavements to form the limestone needs to be pure otherwise large amounts of insoluble residue will mantle the bedrock; it needs to be resistant to frost attack; it needs to be well bedded; and it must not be too soft or highly fractured. For this reason limestone pavements are restricted to portions of the Carboniferous Limestone – they are absent from Jurassic and Cretaceous limestones.

In the post-glacial era, pavement surfaces have been lowered by erosion by around 0.4 of a metre. Eventually, as the valley glaciers and ice lobes encountered shallower gradients, so the great

The mouth of one of the potholes at Buttertubs Pass

masses of material – shattered fragments of rock and boulders which became rounded as they were trundled along by the glaciers, and other rock fragments now ground as fine as flour – were dumped. Sometimes the debris was smeared across the landscape as a 'till plain' of stiff boulder clay, while other masses of material, bulldozed along by the snouts of the glacier, or shed as the glacier receded, were abandoned as hummocky mounds of moraine. The existence of deposits of till that are many metres thick is most obviously apparent at the coast; at places like Flamborough Head, the vertical chalk cliffs are seen to be surmounted by clay weathered into a convex profile, while at resorts near Whitby, like Runswick Bay, a dark-red till containing large rock fragments backs the beach. Yorkshire also has good examples of drumlin fields, the blister-shaped hills being moulded subglacially by slowly moving glaciers from thick beds of till. There is a major drumlin field in Wensleydale, and the northern Pennines are almost encircled by belts of these hillocks.

As the ice sheets and glaciers waned, so torrents of meltwater were released, washing over the glacial materials and depositing sheets and fans of outwash sands. Sometimes the meltwater streams flowed within or under the ice, and eskers, sinuous ridges of sand and gravel, still chart the courses of some ice-bound rivers. The much-visited limestone gorge of Gordale Scar is now regarded as having been cut by meltwater; the Yorkshire Geological Society's field guide notes that: 'The spectacular gorge in the Malham Formation was probably carved as a meltwater channel beneath the Devensian ice-sheet. Upstream of the waterfall the gorge dog-legs abruptly to the northeast, where a series of minor extensional faults have controlled the erosive path of the meltwater channel.' In parts of the Pennines, lakes of meltwater were trapped by the ice which still dammed the upland valleys, and notches in the watersheds, with channels below, can indicate the places where the water escaped; in the Cleveland Hills, Scarth Nick is a channel cut by escaping meltwater which was later used by cattle drovers as a route from the lowlands to the upland grazings. Although several lakes existed in the Vale of Pickering in late glacial and Mesolithic times, the notion of a gigantic glacial 'Lake Pickering' is no longer supported. Meltwater channels could be cut along the margins and beneath downwasting ice; the ice was not a watertight barrier and the existence of channels does not prove the existence of lakes. Smaller meltwater lakes probably did exist, like Lake Scugdale to the north of the Cleveland Hills, formed when ice plugged the mouth of Scugdale

The dramatic cliff of Kilnsey Crag, undercut by the Wharfedale glaciers to produce the daunting overhang. Note the stream at the cliff base, fed by water percolating through the limestone

and meltwater accumulated and escaped to the south-west by cutting a notch and an escape channel through Stony Ridge. The melting of the ice that was responsible for damming Lake Scugdale was associated with the deposition of morainic and outwash material at the foot of the Cleveland scarp. Ice marginal channels, cut by meltwater torrents flowing along the margins of bodies of ice, are numerous around the edges of the North York Moors.

When the ice had melted and the meltwaters had drained into the sea, Yorkshire would have existed as an almost lifeless landscape of barren, soil-stripped uplands and bare plains of outwash sands and till. The pre-glacial drainage patterns had been disrupted and the rivers of the lowlands developed new courses amongst the mounds of debris; in the dales, lakes developed behind the moraines which marked stages in the retreat of the valley glaciers. Accumulations of humus were lost, either scoured away by the advancing ice or buried by till or sand, and this must have delayed

the recolonisation of the landscape by the more fastidious forms of plant life. The discussion of the human relationship with the landscape is the concern of the remainder of this book.

There are many facets to this relationship; physical phenomena could exist variously as resources, influences and constraints. For more than half of the period of post-glacial habitation in Yorkshire, human life has been overwhelmingly dominated by agricultural endeavours. The significance of the upland – lowland division has already been recognised, but past interpretations tended to be wedded to an excessively deterministic view of the limitations imposed upon primitive agriculture by upland conditions – as well as being unduly influenced by the assumption that the North, and its uplands in particular, was an area of pastoral farming. Fowler writes that: 'The traditional view of a Highland Zone dominated by, if not exclusively devoted to, pastoralism is not borne out by the archaeological or indeed palaeobotanical evidence in our period.' Evidence of arable farming in the uplands can be recognised from at least the late Neolithic to the Roman period, with the cairnfields of the North York Moors, which can contain up to 800 cairns of gathered stone, probably being related to episodes of agricultural colonisation. Fowler notes:

The positive evidence from pollen analysis, indicating extensive clearance of alder and hazel scrub 'on an unprecedented scale', rather than the negative evidence of very infrequent burial evidence beneath the cairns, suggests that these remains represent crop production rather than a pastoral economy, and that the cairns are primarily clearance-heaps rather than burial mounds . . . . pollen analytical and C-14 estimates strongly suggest a general clearance of upland wood and scrub for arable farming in the second half of the second millennium BC. That the individual cleared areas are often so extensive and, never having regenerated, are now wilderness landscapes could indicate intensive but shifting exploitation, terminated during or possibly before the onset of the Sub-Atlantic climatic deterioration.

Even the more cyclonic climate of the Iron Age did not prompt an abandonment of arable farming, and the excavators of a farming settlement at Roxby, on the eastern margins of the moors, which was shown to have had a mixed-farming economy, criticised the assumption – deriving from Sir Mortimer Wheeler – that people in the northern uplands had existed upon a diet of 'unmitigated mutton'. Fowler has also criticised the deeply rooted view that prehistoric farming was confined to light soils like those of the Wolds and comparable downlands in the south: 'One of the few certainties is that "the farmers favoured the light soils" model is a gross oversimplification, certainly by the second millennium; another is that the present-day distribution of soil types does not accurately reflect the distribution of basically the same type ranges in c. 2000 BC'. (The latter point relates to the claim that many soils have deteriorated as a consequence of human exploitation.) The areas which resisted prehistoric – and later – colonisation were not so much the uplands as the heaviest or most waterlogged soils of the valleys and plains, the exploitation of which required the attainment of special expertise in land drainage:

Undrained, swampy valleys and heavily forested claylands, could not necessarily be exploited for farming – perhaps there was no need

to – so, in a sense, a sort of balance was probably maintained between food potential and human need. It was later on in the first millennium and in the Roman period that the need arose for more extensive and more intense exploitation of the soil and, then, improved technology made available a greater range of exploitable landscape by releasing some of the hitherto unrealized potential of 'less favourable' soils.

Land quality involves many factors other than altitude and topography, and geological characteristics have been significant. The Carboniferous limestone of the Pennines has numerous distinctive qualities. Because it is both alkaline and riven by a multitude of small fractures, rainfall – which becomes a very mild form of carbonic acid as it falls through the atmosphere and becomes a dilute humic acid as it seeps through the topsoil – is able to penetrate the joint networks and enlarge them by chemical processes. In this way, complexes of caverns formed, some accommodating subterranean rivers; though caves were not indispensable to Upper Palaeolithic and early Mesolithic life, there is ample evidence that the shelter afforded by caves in the Pennines was exploited in Mesolithic and later times. From the agricultural standpoint, the presence of the limestone resulted in the existence of a relatively dry upland landscape in places receiving relatively high levels of precipitation. Within the same locality, one can often contrast the dry, treeless pastures developed on the limestone with the peatbog, cotton grass and heather moor seen on adjacent land floored in Millstone Grit. Raistrick (1968) wrote that:

The high quantity of lime keeps a limestone soil light and crumbly, and this, together with the free drainage provided by the joints, keeps the limestone soil comparatively dry, even in periods of heavy rainfall. It is because of this that there is such a concentration of prehistoric sites to be seen on the limestone terraces; the areas were sought out for their dryness, and there has been almost no accumulation of soil to bury the structures built upon them. The Millstone Grit carries a soil which is a complete

One of the beautiful falls at Aysgarth. The Ure boasts a finer collection of falls or 'forces' than any other British river

contrast with that of the limestone. The coarse sandstones and abundant shales together form a soil that is sandy, porous, and of a low fertility on the grits, and a heavy clay on the shales . . . The Yoredales, with the limestones, shales and sandstones, offer the most variation. The limestone terraces and scars form belts of limestone pasture while the shales and sandstones make a following zone of wetter, sourer soil, so that on a Yoredale hillside these soils follow one another in alternate strips. On the lower slopes the shale bed is often clothed with woodland . . .

These comments are essentially true, though it is noteworthy that the density of the agricultural population on the Millstone Grit has been underestimated, partly because of problems of detection, and it is only recently that the aerial photographic techniques for identifying fields and settlements in the gritstone localities have been explored. The soils derived from glacial drift, which cover parts of the dales and much of the Vale of York, are generally deep and vary in their composition, some of them being friable, while others are cold, stiff clays. One would expect that at times of climate deterioration, such as occurred in the fourteenth century, farmers on the heavier clays, which were prone to waterlogging and were slow to warm in the spring, would have converted from cereal cultivation to livestock farming. In numerous places, the incorporation of drift derived from the limestone of the Pennines results in soils that are sweeter than the acidic bedrock beneath. On the North York Moors, the contrasts encountered between the limestone and sandstone soils of the Pennines are roughly replicated. In the south, variation between the layers of calcareous grit and limestone that compose the Tabular Hills resulted in alternations between pasture and arable farming in the traditional countrysides, while a band of Oxford clay at the foot of the limestone scarp is associated with

One of the bizarre gritstone formations at Brimham Rocks, a National Trust property

good pasture. To the north, there is the poor grazing and heather moor associated with the sandstone. Of the soils of the coastal strip, Atherden and Simmons write: 'All along the coast, the [Scandinavian] ice dropped its boulder clay on top of the Jurassic rocks . . . The boulder clay is a basis of valuable farmland today but causes problems at the coast, where it is prone to slipping.' In the uplands, limestone was generally found in reasonable proximity to the acid soils associated with the sandstones, though Raistrick (1991) suggests that it was not until the sixteenth century, many centuries after the first burning of lime in kilns for use in building works, that the value of lime as a means of improving sour land was realised. He notes that: 'In the limestone areas of the Pennines large numbers of farmers and landowners built small kilns for the improvement of their own uplands, while on the flanks of the Pennines and the gritstone areas, a demand for lime was created which encouraged the building of larger kilns as a commercial adventure, producing lime for sale.'

The significance of geological contrasts could be intensified at times of environmental crisis. Major episodes of climatic deterioration are associated with the start of the later Bronze Age and the fourteenth century. There was a major exodus from the British uplands during the Bronze Age, caused by the onset of cool, moist, cyclonic conditions which have been associated by some with violent volcanic eruptions on Iceland. Although all marginal communities were affected, the problems must have been particularly severe on the coarse, leached, calcium-deficient soils of the sandstone uplands, where waterlogging, acidification and the formation of peatbog proceeded rapidly. During the twelfth and thirteenth centuries, population pressures caused a colonisation of marginal agricultural environments in both upland and lowland localities, with the clearing of low-grade land which had usefully served as woodland. Such environments were least able to sustain agriculture when the climate deteriorated and were most susceptible to the adverse consequences of over-cropping and inadequate fallowing or manuring. The heaviest clays and the more infertile and sandy soils will have been the first to be deserted. Natural catastrophes could also occur in a more

immediate and dramatic form. From Flamborough Head, a low and vulnerable coast extended southwards to Spurn Head. The coast as it existed prior to the Pleistocene glaciation roughly followed the eastern foot of the Yorkshire Wolds; subsequently, a new coastline was formed as the rising sea eroded the till deposited by the ice sheets, and in the Holderness area the rapid and continuous advance of the sea has had severe consequences. Currently, the cliff line of vulnerable till is receding by up to 2 metres each year. It is estimated that, since Roman times, a coastal land strip some 35 miles (56km) long and 1–3 miles (1.6–5.8km) wide has been lost to coastal erosion; between Bridlington and Spurn Head, the sea washes over land which once supported around thirty villages and a section of the old Beverley to Bridlington road. Old Bridlington, Auburn, Hornsea Beck, Colden Parva, Old Aldbrough, Owthorne, Old Withernsea, Out Newton and Old Kilnsea are amongst the settlements known to have been lost. Mud, washed into the mouth of the Humber by the tides, and debris, swept down from the eroding cliffs to the north, combined to build the slender sweep of Spurn Head. In this section of the coast, old sea banks, walls, groynes and plantations of marram grass chart some of the numerous efforts to reclaim and secure land from the sea. Settlements lost during the medieval climatic crisis, when sea storms increased in frequency and intensity, included not only villages but also the significant trading town of Ravenser-Odd, which lies submerged off Spurn Head. In 1355, bodies were washed from their graves in the churchyard and several inundations of the town followed; in 1361, the Abbot of the nearby Cistercian abbey of Meaux berated the evil sea: 'by its wicked works and piracies it provoketh the wrath of God against itself beyond measure'. The spit has migrated westwards, leaving Ravenser-Odd and its sister town of Ravenser behind. It is also gradually lengthening, and is estimated to have grown by some 2,313 metres between 1676 and 1851.

Resources other than farmland influenced the colonisation of Yorkshire. The availability of flint, which occurs as nodules within the chalk, is restricted to the Wolds and the adjacent coast, though the numerous flint scatters in the northern

and western uplands reveal an active trade in this essential commodity. During the Neolithic period, flint was systematically exploited at the coast, with nodules from the best seams being collected from the beach as they eroded out of the chalk and knapped to produce a variety of axes, knives

A spectacular wave-cut arch in the chalk cliffs near Bempton. In time, the roof of the arch will collapse, leaving a stranded chalk pillar or needle, but a new arch will form as the waves enlarge the cave seen just landward of the arch

and scrapers at cliff-top or inland sites. Deposits of waste material can be detected at cliff-top sites around Flamborough Head. Axes of stone were also valued, the best being imported in substantial numbers from axe factories exploiting volcanic tuff at Pike O'Stickle and other sources of this rock in the Langdale Pikes in Cumbria. During the Bronze Age, battleaxes made from the dolerite of the Whin Sill were imported. The Millstone Grit was of value to agricultural societies for the grinding of grain, not only by millstones, but also in the querns or hand-mills associated with farming settlements from the prehistoric to the medieval periods. Until the development of pollen analysis, the numerous discoveries of querns provided the best evidence of arable farming in the uplands. Raistrick (1991) notes that sandstone 'bakestones', made from a fire-resistant sandstone and used to build ovens or as slabs for oatcake baking, were of importance from prehistoric to relatively recent times. More than twenty places perpetuate the name, including Bakestone Fell, where the stones were quarried. Helwith flags, quarried from Silurian rocks in Ribblesdale, were used for

domestic flooring and kitchen and pantry shelving, as well as for making rainwater storage tanks. Although vernacular architecture was widely accomplished in timber, wattle and daub in Yorkshire until the seventeenth or eighteenth century, the region is well-endowed with building stones of many types. During the medieval period, a group of quarries around Tadcaster were exploited to extract limestone of the finest quality for use in major church-building projects. Lesser undertakings used rubble or freestone of various types, with Jurassic limestone being employed in the settlements around the southern margins of the North York Moors, and Millstone Grit and other sandstones being used in the Dales. Raistrick (1991) remarks that:

> In most of the Yorkshire Pennines the Millstone Grit series forms the upper part or cap of the fells and few villages are far removed from a bed of grit. In Dales building, therefore, it was a common practice to open a small quarry on the waste of the moorland, at the lowest of the suitable grits, get the stone and trim it on the spot, then sledge it down the hillside to where it was wanted.

Millstone Grit was also exported, with its resistance to attack by sea water making it a preferred

Bishopdale Beck is a 'misfit stream', its mighty valley having been carved largely by glaciers diverted by the bulk of Buckden Pike

Gormire, a hollow in the Hambleton Hills escarpment, caused by a mud flow which dammed a glacial meltwater channel

choice for various harbour-building operations; the Bramley Fall stone, quarried near Leeds, was a Millstone Grit of high quality. Sandstones from the Yoredale Beds were used by small-scale operations to produce building stones and roofing 'slates', while larger quarries worked the Elland flags, Oakenshaw rock and other sandstones from the lower coal measures. Westwards from Elland to the moors beyond Huddersfield, a chain of quarries exploited the Carboniferous sandstones. Quarries in the Halifax locality exported flags to pave the streets of several cities, and Jurassic sandstone from the North York Moors paved streets in London. Not only was stone used in the post-medieval centuries for building a spectrum of dwelling types, but it was also used very frequently for roofing the houses. This required special fissile sandstones which, after weathering in the wet and frost, could be split into sheets around two to three centimetres in thickness. The best materials were found in the Yoredales and extracted from small quarries like the one at Walden Head, above Starbotton in Wharfedale, or at Blackstone Edge, by the Lancashire border. Almost all accessible stones were used for drystone walling, except the soft, frost-prone chalk, which was used for some enclosures within villages in the Wolds, but not for field walling. Since the facilities to transport bulky, low-value consignments of stone did not exist until recently, walling

stone was obtained locally from small quarries, and the walls provide good indications of the local geology – with the 'magpie walls' incorporating both Carboniferous limestone and Millstone Grit marking the geological boundary between the two.

Yorkshire was deficient in the copper and tin which rose to great significance in the Bronze Age; there are a few small sources of copper in and around the Pennines and, further afield, in larger reserves at Ecton, near Leek; copper is known to have been worked in prehistoric times at Alderley Edge, in Cheshire. Iron, from various sources, is relatively widespread; it is found as clay ironstone in the mudstones of the Coal Measures. Iron-rich deposits that formed on the Lias of the North York Moors were mined near Grosmont and gave rise to the Cleveland iron industry. Veins of lead are found in the Lower Carboniferous rock and were exploited on a large scale during the eighteenth century in Swaledale and in the Greenhow Hill locality, above Pateley Bridge, where lead was worked in Roman times. The physical environment provided certain opportunities and constraints, but it would be impossible to deduce the exact manner in which it may have influenced human development in Yorkshire or elsewhere. Some causal relationships are clear-cut; West Yorkshire's involvement and transformation in the steam-powered phase of the Industrial Revolution was plainly conditioned by the presence there of substantial quantities of coal. However, other presumed relationships can be illusory; for instance, with the exception of a few simple examples, like bridging-point settlements, it is not easy to relate village locations to the convergence of favourable aspects within their local environments.

REFERENCES

M. A. Atherden and I. G. Simmons, 'The landscape' in D. A. Spratt and B. J. D. Harrison (eds), *The North York Moors* (Newton Abbot: 1989).

J. Boardman (ed.), *Field Guide to the Periglacial Landforms of Northern England* (Cambridge: 1985).

S. Ellis (ed.), *East Yorkshire Field Guide* (Cambridge: 1987).

A. Fleming, 'The genesis of pastoralism in European prehistory', *World Archaeology,* 4(1972), pp.179–91.

P. J. Fowler, *The Farming of Prehistoric Britain* (Cambridge: 1983).

Sir Cyril Fox, *The Personality of Britain: Its Influence on Inhabitants and Invaders in Prehistoric and Early Historic Times* (Cardiff: 1932).

H. S. Goldie, 'Human influence on landforms: the case of limestone pavements', in K.Paterson and M. M. Sweeting (eds), *New Directions in Karst* (Norwich: 1986).

A. Goudie, *The Landforms of England and Wales* (Oxford: 1990).

P. E. Kent, *Eastern England from the Tees to the Wash* (London: 1980).

C. A. M. King, *The Geomorphology of the British Isles: Northern England* (London: 1976).

D. Lewis (ed.), *The Yorkshire Coast* (Beverley: 1991).

NERC , *British Regional Geology: The Pennines and Adjacent Areas,* 3rd edn (London: 1978).

E. J. Pounder, *Classic Landforms of the Northern Dales* (Sheffield: 1989).

A. Raistrick, *The Pennine Dales* (London: 1968).

A. Raistrick, *West Riding of Yorkshire* (London: 1970)

A. Raistrick, *Arthur Raistrick's Yorkshire Dales* (Clapham: 1991).

CHAPTER TWO

The Pioneers' Landscape:
The Mesolithic

ONE WILL NEVER KNOW WHO THE FIRST Yorkshireman or woman was. He or she would certainly have been a member of a hunting and foraging band which wandered northwards in pursuit of game during one or other of the long, mild interglacial periods. Whether this person belonged to the *Homo erectus* line, (which is represented elsewhere by the famous fossils of Peking and Java man and has recently been discovered in England at Boxgrove), one cannot tell. Nor can we know if he or she was a closer relative of modern humanity of the Neanderthal type – or perhaps early ancestor in our own *Homo sapiens* lineage. Trapped beneath an accumulation of calcite on some cave floor or else buried beneath metres of glacial drift may be the partly fossilised teeth that could answer the question, though it is likely to be a very long time before they are found. Whatever the answer may be, any visible trace of such pioneers would have been obliterated by the ice and debris of the next glaciation to occur, so that, for us, the saga of humans in Yorkshire effectively began anew in the closing phases of the last great glaciation.

The maximum extent of ice in Britain occurred around 18,000 radiocarbon years ago, and most authorities agree that there was no significant human population of hunters in Britain or elsewhere in north-western Europe at this time. This is not to say that Yorkshire was an absolute desert during the last ice age. There are summers even during glaciations, and the prehistorian Alan Turner has argued that during the summers of the last ice age, herds of reindeer would migrate into northern England, with bands of resourceful human hunters appearing though in a less than dominant role, amongst the various carnivores that preyed upon them. He demonstrated the evidence for the seasonal presence of deer, horses, hyenas and bears and wrote that:

> Of course seasonal presence of animal species does not provide direct evidence of human movement patterns, but has nonetheless a direct relevance to the problem. If two major prey species and a major predator made annual migrations, then there can be little doubt that man did so too. Evidence of the seasonality emphasises that events recorded in Britain were simply one part of the annual round of activities of the men and animals represented.

It is always easier to underestimate rather than overestimate the achievements of early human communities, so it is at least possible that the hunters' relationship with the deer was evolving towards that of herdsmen who culled their herds rather than of hunters who ranged widely in the hope of finding game. However, the understanding of human life in Yorkshire around the transition from the Palaeolithic to the Mesolithic is undermined by the paucity of evidence. Not only will the humans have been few in number and, quite probably, migratory in habit, but the gap in time that separates their present from ours is so wide that the material evidence for their existence is likely to have been removed or obscured as rivers changed course, soil was washed from slopes and streams deposited their burdens of silt. Much of the best evidence for Mesolithic life derives from the preservation of ancient pollen grains, allowing the reconstruction of past vegetation patterns. In fact, plenty of pollen has survived from the late glacial period and there is an abundance from the early Mesolithic period. It reveals that although trees were few and were confined to sheltered positions, much of the land was grassland comparable to some modern tundra plains.

By around 13,000 years ago, the climate of Yorkshire had warmed significantly, although it was still cool by present standards and could be compared to the conditions which now prevail in Lapland, with very cold winters and tepid summers. As a hunter of the herds of ungulates, man would have been familiar with beasts like the horse, reindeer, red deer and European bison, as well as the red and Arctic fox and the hare, and would have been wary of dangerous competing predators like the wolf and the bear. (Wolverine Cave at Stump Cross Caverns, near Grassington, is named after the vicious carnivore which occupied the cave more than 100,000 years ago, whose descendants were still present in the North around 12,000 years ago and whose skull is displayed in the visitor centre.) The landscape that the hunting bands explored was colonised by tundra vegetation, with dwarfed forms of the hardy birch and willow, reindeer moss and alpine flowering plants producing scenes not unlike those now found in parts of Alaska, Iceland and northern Scandinavia. The conventional late glacial chronology is: *c.* 26,000–13,00 BP Dimlington stadial

Mesolithic hunters visited the Kirkdale caves, here exposed in the face of an abandoned limestone quarry. Dr William Buckland explored the caves in 1821 and discovered the remains of around 300 hyenas and bones of elephants, bison, lions, bears, rhinos, boar, horses and wolves. The photograph shows one of the two openings of the linked cave system

(coldest phase); 13,000–11,000 BP Windermere interstadial (warmer phase); 11,000–10,000 BP Loch Lomond stadial (colder phase).

Between about 11,000 and 10,000 BC, the temperate conditions were negated by a sudden reversion to near glacial conditions, which exterminated the birch woodland colonising the wilderness and brought a return to tundra vegetation. This severe cold period was followed by a sustained warming, so that by about 10,000 years ago the last of the snow patches which lingered in the shaded hollows of the Pennines had melted and thereafter, the Yorkshire climate seems to have become milder than that of today.

In popular mythology, the ancient hunting communities tend to be characterised as 'cavemen'. Of course, one can only be a caveperson if one has a cave to live in – and away from the cavern-pocked limestone areas, caves are very few and far between. Where convenient caverns and rock shelters existed, they probably served only as seasonal abodes to be occupied when the migrating herds of horse and deer drew the hunters into limestone country. Elsewhere, the bands must have occupied open sites, probably building tents of hide, with boulder walls to serve as wind-breaks, or, perhaps, igloos. The limestone areas of the Pennines contain a number of caves which are known to have experienced prehistoric occupation at various periods, although disturbances to the stratigraphy of the cave-floor deposits tend to make dating attempts rather difficult. Nomadic hunters lacking beasts of burden or wheeled vehicles must travel light, and their only possessions would have been their tools of stone, bone or antler – needed as weapons, for

The narrow, tunnel-like interior of one of the Kirkdale caves

butchery, tailoring, digging and tool-making – and their warm fur and hide garments. Frequently, cave excavations reveal the bones of long-lost animals – hippo, lion and elephant from inter-glacials, and mammoth, horse, ox and deer. Without clear evidence of human artefacts or butchery, one can never be sure whether these are human hunting spoils, the remains of prey dragged into the cave by bears or hyenas, or bones washed in by flood waters. It is likely that a number of caves, like the Queen Victoria cave above Settle, experienced intervals of human occupation during interglacials or even during mild interludes within glaciations, as well as harbouring folk who penetrated northwards at the close of the last ice age. Because Yorkshire lies in the chillier north of England, the heritage of Palaeolithic relics is more modest than that associated with more southerly parts of the country, which could have experienced many centuries of Upper Palaeolithic occupation while Yorkshire remained too frigid to sustain human life. Even so, it is not so far removed from the cave complex at Creswell Crags, on the Nottinghamshire/Derbyshire borders, where a rich assemblage of interglacial and other Palaeolithic remains have been discovered.

The Mesolithic or middle Stone Age period bridges the broad gap between the decay of the last ice age – along with that of the Palaeolithic lifestyle, which was based upon hunting the large herbivores of the open tundra – and the first appearance of farming in England. This innovation arrived after about 5000 BC, when agriculture had diffused north-westwards across Europe to reach the shores of the English Channel. The environmental consequences of the warming of the climate that created the Mesolithic period were profound. As Barnes has described:

> The crucial effect of these climatic changes was the replacement of open tundra by forests, ultimately to considerable altitudes in the uplands. Changes in food supplies resulted in radical changes in the types of animal population. Open country forms disappeared, apparently abruptly in most cases, and were replaced by woodland species such as red and roe deer, elk, aurochs and wild pig . . . the biomass (the

overall density of the animal population) in forested areas is much less than that of open environments, and . . . forest species are also less gregarious in their habits. Thus the easily culled herds of reindeer, for example, were replaced by a more diffuse population red deer and aurochs in particular, animals which were far less receptive to man's presence and thus less easy to exploit.

The communities occupying Yorkshire around the beginning of the Mesolithic period were obliged to reorganise their lives to adjust to the sweeping environmental changes brought about by the warming of the climate and the gradual recolonisation of the countryside by broadleaved woodland. As a consequence of the great, and presumably traumatic, cultural shifts required for survival in a changed environment, Mesolithic communities became essentially people of the forest, valley and lakeshore – while they also ranged across the more open hunting grounds of the upland plateaux (though the pollen evidence suggests that, in Yorkshire, nearly all such plateaux were forested).

Although the vast herds of large tundra herbivores had been displaced, to be replaced by the more elusive woodland creatures, the woods now offered a varied diet of fruits, shoots, nuts and roots, and the rivers and lakes again abounded with fish. Hazel was prominent in the vanguard of deciduous trees and its nuts appear to have been especially valued as a source of nourishment that could be gathered in autumn and stored to offset the shortages of winter. The lost and the new lifestyles both required deep reserves of resolution and ingenuity for survival. The Mesolithic year involved a juxtaposition of times of plenty, like autumn, with its fruit, nuts, fungi and the presence of inexperienced young prey, and winter, when the vegetable resources vanished, daylight for hunting was reduced, and communities may have been drawn to the more reliable resources of the beach, lake and river (though this is now debated). One can gain a few insights into the abilities of the Mesolithic people by inspecting their artefacts: minute slivers of flint used to tooth and barb tools and weapons, skilfully fashioned fish-hooks of bone, and lethally barbed harpoons of antler. Since the woodland creatures were fewer, smaller and much more difficult to hunt, the change in conditions must have made heavy demands on human adaptability. Communities whose ancestors had lived by culling just one or two large herbivore species, like the reindeer and the horse, had been obliged to develop survival strategies which involved exploiting almost every food opportunity that the environment provided. Fish and shellfish, wildfowl, rodents, crustaceans, roots and shoots, nuts and berries – nothing could be ignored. This demanded a broad awareness of the contents of each ecosystem and an enhanced sensibility to season and place. But at least the return of vegetable foodstuffs must have been a considerable bonus, and many of the tools edged with rows of tiny flint microliths which have been regarded as weapons might just as easily have been used for scraping and slicing vegetable matter.

Some vestiges of cave life continued in the Pennines, and Mesolithic relics have been found in Calf Hole cave, Skyrethorns, and the Kinsey and Attermire caves near Settle, as well as at Queen Victoria cave, nearby, where a bone harpoon was discovered. In 1821, by Hodge Beck, near Kirk Dale, on the southern edge of the North York Moors National Park, quarrymen discovered a cave which contained bones of a range of animals which had either used or had been dragged into the cavern at various glacial and interglacial times. They included lions, rhinoceroses, and mammoths, while masses of hyena remains were also found. The delicately worked flints left by Mesolithic people were there, too, hunters who may have had their bases by the coast and have made summer hunting forays into this area. Dozens of open Mesolithic camp sites have been found on the high ground of the North York Moors and Pennines – and cave life was a minority taste, although many caves continued to be visited until there was a severe deterioration of the moorland soils during the Bronze Age.

The best example of a Mesolithic settlement, and an archaeological site of international importance, was established by the side of a former lake at Star Carr in the Vale of Pickering. This site was inhabited at quite an early stage in the period, but already the pioneering birch forest was being overshadowed in places by a rich deciduous cover of hazel, elm, lime and oak. A recent reinvestigation of the dating evidence by Day and Mellars has revealed that the occupation of the site began over 1,000 years earlier than the uncalibrated carbon-14 dates had implied. They suggested that there were two phases of human occupation here: the first beginning about 10,700 years ago; the second at 10,550 years ago. Each period of occupation lasted for several decades, during which time the vegetation in the vicinity of the site was regularly burned. Writing in the early 1950s, the original excavator of the site, J. G. D. Clarke, identified what he regarded as a temporary camp that was used in the winter – spring period. It was deduced that families would gather here when conditions on the higher ground were harsh, and depart for the open uplands again when the warmth returned. A more recent reanalysis of the large mammal remains that were found there by Legge and Rowley-Conwy caused them to conclude that the site was not a base camp, but a hunting camp to which short visits were made, mainly in the early summer, when the hunting was easy – though they thought that the occupation of the site could have extended beyond the May–September period. Schadla-Hall has questioned whether there would have been a need to migrate from sites like Star Carr, which was in an ecologically varied locality, rich in the plant and animal resources needed to support a year-round occupation. There is also doubt as to whether the North York Moors would really have been too cold for the red deer in winter – but even if the deer did winter in the lowlands and migrate to the uplands, there were ample fish resources to support a lowland community during the summer.

We now have a more detailed understanding of the setting of Mesolithic life at Star Carr. Cloutman and Smith have described how, at the time of the occupations, the inshore reed swamp composed of the great reed had largely been replaced by a fen environment dominated by the saw-sedge. At the margin of the fen, there were ferns and sedges; when the human occupation was over, the saw-sedge grew over the platform that had been built at the edge of the lake, and willow became established there, too. The site chosen for occupation was on the northern side of a narrow outflow channel from a reasonably large lake, one of several then existing in the Vale of Pickering; it covered both dry land and wet ground and it lay at the lower end of a shallow gully which ran down into the willows and sedges from the birch-clad hillocks to the north. The dampest ground was consolidated by birch timbers, and the platform of brushwood, stones, clay and moss built here provided a living and working area where the debris of Mesolithic life accumulated. It seems that the site was chosen for its shelter and for its easy access to the lake, since it was there that the open water came closest to the shore.

A very rich assemblage of animal material, mainly the debris from countless meals, survived to be identified. So we know that the Star Carr community shared the countryside with wild pig, beaver, wolf, fox, pine marten, badger, hedgehog – most of these being woodland creatures – and also a wide variety of wildfowl. The five main animals to be exploited for food by the hunters of Star Carr were the aurochs, elk, red deer, roe deer and wild boar. One particularly interesting find, which has inspired controversy ever since, comprised stag 'frontlets' – part of the skull of the stag with the stumps of the antlers attached. It is suggested that this was used as a mask in a hunting ritual. Another skull found here could represent an early stage in the domestication of the dog from its wolf forebears. Star Carr was not the only site to be occupied in the Vale of Pickering during the Mesolithic period. The lake offered diverse resources, and Schadla-Hall has described how many thousands of worked flints have been found at Seamer Carr, at the former junction of dry land, fen carr, reed swamp and open water.

The evidence in the Pennines is much less comprehensive, consisting almost entirely of scatters of worked flint. Even so, by the mid-1970s almost 540 flint sites had been recorded in the central and southern Pennines above the 366 metres contour – so this was hardly a neglected wilderness. Careful study of the flint-working sites suggest that they were associated with encampments and hearths; often the flints occur in a few adjacent concentrations, hinting at settlements of several temporary dwellings or

wind-breaks, each harbouring a hunter who spent the evenings chipping out blades to tip an arrow or barb a spear. Such hunters would, according to the traditional view, follow the game up to the high plateaux in the summer, returning to the valleys, plains and lakeshores and strand-lines in winter to fish and consume the stores of hazelnuts which had been gathered during the autumn while venturing out to hunt the red deer which wintered in the sheltered woodlands.

When these communities first penetrated the Pennines, the uplands would have been unforested, but as the climate rapidly ameliorated, so hazel scrub appeared amongst the grasslands and hazel then combined with birch, stands of pine, and with the oak growing on the more sheltered slopes. By the climatic optimum of around 7000 BC, the tree-line in the Pennines probably extended right up to about 720 metres, though with the woodland cover being light, some late glacial tundra species survived on a few higher summits and ridges, as in upper Teesdale. Before the woodland could establish a hold on some extensive areas of the uplands however, man intervened, and undergrowth was burned to prevent the forest regenerating and to preserve grazing areas with grass and soft browse, making it easier to control the movements of the deer and establishing open killing grounds. Whether humans may thus have created an artificial environment above the 366 metre contour, with trees even being burned on rotation every 5 to 15 years, is debatable, though the evidence of open country is suggested by the concentration of the later Mesolithic sites in the Pennines in the 366- to 450 metre zone. Hayfield *et al.* have described how Mesolithic flint scatters are often associated with upland water supply sites; they mention Vessey Ponds, high in the chalk Wolds, as well as sites at Marsden and Nab Water in the West Riding. Perhaps these were the places were hunters waited to ambush animals as they came to drink – ponds and springs would always have attracted camp sites, though the attraction of ponds in the dry Wolds will have been far greater than in the Pennines. On the upland plateaux of the North York Moors, the scatters of microliths are also very numerous.

A small site from the earlier part of the

Mesolithic was explored by Gilks. It lies on the eastern edge of the Pennines, at Nab Water on Oxenhope Moor. It was probably occupied in summer and autumn, serving as a base camp and hunting camp for a small community engaged in exploiting the plant and animal resources of the heathlands and woodland of the locality. Around 8000 BC, the hill slopes there were probably covered in grass, herbs and heathland, with clumps of birch and a little pine punctuating the landscape. Five hundred or so years later, the pine began to expand; around 7200 BC, a hazel scrub with birch and pine was becoming the dominant vegetation, interspersed in places with alder, oak, lime and elm.

The environment varied from region to region during the Mesolithic, as at every other time. It was assumed that woodland had colonised the dry chalk of the Wolds, but this was disputed by Bush, who suggests that grassland carpeted the chalk when woodland was expanding elsewhere

As the Mesolithic climate warmed and hardy deciduous trees colonised the countryside, much of Yorkshire will have resembled this vista of birch wood at Strid Woods in Wharfedale

in Yorkshire, and that human activity then helped to sustain this grassland, though when the turf was subsequently removed for cultivation, there was severe soil erosion. Of the North York Moors, Atherden and Simmons write:

The whole upland was covered in oak forest in about 8000 BC except for some areas where bog growth had started and perhaps a few exposed hill-tops and ridges. It was the use of the forests for construction, fuel and grazing, coupled with the management of the open landscape to prevent tree regrowth (a process which started in 8000 BC) which produced the moors.

In time, the removal of the natural woodland from the uplands of Yorkshire resulted, in places, in the saturation and impoverishment of soils and in the formation of peatbogs. It encouraged the establishment and expansion of the heather moors which were destined to be regarded as a most characteristic facet of the 'natural' scenery of Yorkshire. In the northern Pennines and the North York Moors, we now know that the wear and tear on the environment resulting from the clearance of woodland depleted the nutrients in the soil and caused elm, ash and lime to retreat from many areas. These events are of more than an antiquarian interest. They signify that man had ceased to be one hunter among several other animal predators, but had consciously *chosen* to modify his environment: the most significant and far-reaching milestone on the road to civilisation. Significantly, too, this human interference led to the degradation of the setting, a process which is ongoing. If the interpretations of the early days of humankind in Yorkshire are somewhat uncertain and controversial, the deliberate modification of the setting by Mesolithic communities can be recognised as the first step in a chain of human interactions with the landscape which have culminated in the countryside that we see today.

REFERENCES

M. Atherden and I. G. Simmons, 'The landscape' in D. A. Spratt and B. J. D. Harrison (eds), *The North York Moors* (Newton Abbot: 1989), pp. 11–27.

B. Barnes, *Man and the Changing Landscape: A Study of Occupation and Palaeo Environment in the Central Pennines* (Liverpool: 1982).

M. Bell and M. J. C. Walker, *Late Quaternary Environmental Change* (Harlow: 1992).

P. C. Buckland and K. Edwards, 'The longevity of pastoral episodes of clearance activity in pollen: the role of post-occupation grazing' *Journal of Biogeography*, 11 (1984), pp. 243–9.

M. B. Bush, 'Early Mesolithic disturbance: a force on the landscape', *Journal of Archaeological Science*, 14 (1988), pp. 453–64.

S. Caulfied, 'Star Carr – an alternative view', *Irish Archaeological Research Forum*, 54(1978), pp. 15–22.

F. M. Chambers (ed.), *Climate Change and Human Impact on the Landscape* (London: 1993).

J. G. D. Clark, *Excavations at Star Carr* (Cambridge: 1954).

E. W. Cloutman and A. G. Smith, 'Palaeoenvironments in the Vale of Pickering, part 3: environmental history at Star Carr', *Proceedings of the Prehistoric Society*, 54 (1988,) pp. 37–58.

J. Gale and C. O. Hunt, 'The stratigraphy of Kirkhead Cave, an Upper Palaeolithic site in northern England' *Proceedings of the Prehistoric Society*, 51 (1985), pp. 283–304.

J. A. Gilks, 'Earlier Mesolithic sites at Nab Water, Oxenhope Moor, West Yorkshire', *Yorkshire Archaeological Journal*, 66 (1994), pp. 1–19.

C. Hayfield, J. Pouncett and P. Wagner, 'Vessey Ponds: a "prehistoric" water supply in east Yorkshire?' *Proceedings of the Prehistoric Society*, 61 (1995), pp. 393–408.

R. L. Jones and D. H. Keen, *Pleistocene Environments in the British Isles* (London: 1993).

J. A. Legge and P. A. Rowley-Conwy, *Star Carr Revisited* (London: 1988).

P. A. Mellars, 'The Palaeolithic and Mesolithic' in C. Renfrew (ed.), *British Prehistory: A New Outline* (London: 1974), pp. 41–99.

P. A. Mellars (ed.), *The Early Postglacial Settlement of Northern Europe* (London: 1978).

N. Roberts, *The Holocene: An Environmental History* (Harlow: 1992).

R. T. Schadla-Hall, 'The early post glacial in Eastern Yorkshire' in T. G. Manby (ed.), *Archaeology in Easter Yorkshire: Essays in Honour of T. C. M. Brewster FSA* (Sheffield: 1988), pp. 25–34.

I. G. Simmons, M. Atherden, P. R. Cundill, J. B. Innes and R. L. Jones, 'Prehistoric environments', in D. A. Spratt (ed.), *Prehistoric and Roman Archaeology of North East Yorkshire*, British Archaeological Reports, British Series No. 104 (Oxford: 1982).

P. Stonehouse, 'Mesolithic sites on the Pennine watershed' *Greater Manchester Archaeological Journal*, 2 (1987–8), pp. 1–9.

A. Turner, 'Predation and Palaeolithic man in northern England', in G. Barker (ed.), *Prehistoric Communities in Northern England* (Sheffield: 1981), pp. 11–26.

C. T. Williams, *Mesolithic Exploitation Patterns in the Central Pennines*, British Archaeological Reports, British Series, No. 139 (Oxford: 1985)

J. Wymer, *The Palaeolithic Age* (London: 1982).

CHAPTER THREE

Farming and Territory: The Neolithic

YORKSHIRE AT THE DAWNING OF THE NEOLITHIC, almost 7000 years ago, was a land very different from the one that exists today. The surfaces of the higher plateaux might display seemingly familiar moorland, but the valley slopes would be cloaked in forest, many valley floors would be dappled by lakes and marshes and traversed by winding, shifting rivers, while on the plains heavy woodland, spattered here and there with wetlands, would span the horizons. Almost every corner of the landscape would have had its owners and its bounds, however, for already the countryside was almost certain to have been partitioned between different clans, each with its familiar hunting, collecting and grazing territories. Current archaeological opinion seems to be converging upon the view that the countrysides of the late Mesolithic period were more densely populated than had previously been thought and that human survival need not have depended on a perpetual sequence of long cross-country journeys in search of food. A group territory might often embrace lowland wintering grounds beside a river or lake, a zone of woodland with its game and plant resources, and an expanse of deforested upland with open hunting ranges and pastures – although the groups controlling the richest territories might have had little cause to move.

Change of the most revolutionary kind was on its way. Long ago, when people living in Yorkshire were slowly adapting their ways of life to cope with the changes wrought by the warming of the climate and the northern advance of the forest, the rudiments of farming had been discovered in the Near East. By about 8000 BC, Jericho was a substantial agricultural village with mud-brick houses and the glimmerings of civilisation. By 5000 BC, the old English lifestyle of hunting the great herds of reindeer and horses across the open tundra lay far away in the past – and in the meantime, the first Agricultural Revolution had advanced its frontiers through the Balkans, across the Danubian plains, and north-westwards to the shores of Europe. From Belgium to the Ukraine, cereals were being cultivated by the sides of watercourses and cattle seem to have been stall-fed in long timber cowhouses.

Whether it was just the idea of farming that was exported to Britain, or whether the new way of life was transplanted by groups of immigrant settlers, we do not know. Quite probably, both sources were involved to some degree, but, before very long, each community here faced a choice between the old and proven ways and the challenge of the new. Trauma and even conflict may have resulted as different perceptions were brought to bear upon the countrysides, where one community's food supply was now another community's vermin. The new ways triumphed, so that instead of living *off* the land, people now began to live *on* it, bound to the plots and crops where their seeds had been sown. Almost certainly, it was the southern and south-eastern parts of England that first experienced the shocks of the changes, but the notion of the farming lifestyle lapped across the country quite rapidly and had been accepted in Co. Tyrone, far away beyond the Irish Sea, well before 4000 BC. At first, farmers must have had the pick of the light but rich alluvial soils and the dry, lime-rich lands of the chalk uplands, but soon the best plots were taken up and they could be less selective, so that agriculture forged inroads into the poorer terrain of the dales and sandstone uplands.

Farming is regarded as the hallmark of the new Stone Age or Neolithic period. If one looks at a stone axe of the preceding Mesolithic period, one sees a bulky, rather pick-like tool, useful for grubbing up roots or for tasks involving crushing. However, a polished stone axe of the Neolithic period is instantly recognisable for what it is: a remarkably efficient tool for felling or splintering trees (as well as an effective weapon and, in its more grandiose forms, an impressive status or ritual object). Change from Mesolithic hunting and foraging to Neolithic farming was not instant or absolute. Hunting continued on the high moors of the Pennines, and the dainty leaf-shaped arrowheads of the Neolithic period are often found close to the places where one finds the minute and triangular stone arrowtips lost or discarded by the earlier hunters of the Mesolithic. In contrast,

the polished stone axes tend, in the Pennines, to be discovered on the flanks of the uplands and the sides of the valleys, showing that, while hunting endured on the open uplands, plots were being hacked into the broadleaved forest of the slopes below.

The discoveries of polished stone axes in the central Pennines reveal the development of commerce in the Neolithic societies. Prehistorians

The late Neolithic stone circle of Castlerigg in Cumbria. Some experts believe that the Cumbrian stone circles could have served as the market places of the Langdale stone-axe industry – in which case, herdsmen from Yorkshire may have traded here during their movement of cattle to high upland grazings in the west. Other experts envisage a sophisticated land and seaborne export trade in Cumbrian axes, while still others think that the axes were prestigious gifts which were gradually dispersed among the upper echelons of Neolithic society. All these interpretations could be correct, of course

have shown that, from a sample of 54 axes found there whose precise stone types can be indentified, 22 originated in Cumbria, 4 came from the Graig Lwyd axe factory in North Wales, 3 from another famous axe factory in Co. Antrim, 2 from the tough dolerite of Whin Sill (the outcrop which carries Hadrian's Wall), and single examples could be traced to sources in the south-west of England and Wales and the Nuneaton area. In addition to these, there were 20 flint examples, many of which must have had East Anglian or, perhaps, Sussex origins. Quartermaine and Claris have shown that: 'the Cumbrian axe factories were producing axes on a massive scale consistent with their nationwide distribution'. The south scree which runs down from the domed summit of Pike of Stickle in the Langdales was calculated to contain about 450 metric tons of waste flakes from axe-making, leading to the estimation that this represented the debris from the manufacture of 45,000–75,000 axes. Many of these must have been destined for export to Yorkshire. Such evidence must surely dispel any notions that the early farmers belonged to isolated, struggling communities. Plainly, a British system of axe-trading networks had been established, with its own specialised quarrying and manufacturing centres, its trade routes and bartering systems. Just a few of the axes found are unpolished 'rough-outs', hinting that they were sometimes exported in an unfinished form, with the final stages of grinding and polishing being accomplished by the 'purchasers' of these essential and prestigious tools. Also, a few of the axes are of such a size and of such lustrous or finely marked stone that they seem to have been reserved as symbols of high status.

The evidence of axe-trading is apparent in the low plains of the Vale of York, where a flint-based industry seems to have been supplanted by one which used other exotic stones, with those of the Lake District again playing a notable role. One cannot know just how the trade was organised, though in some cases sea transport seems to have been an important means of distributing the axes. Also, Manby has suggested that cattle drovers based in the Vale of York may have moved their herds to the Pennine uplands during the summer, then obtained axes from the Lakeland factories

and carried them back for trade with the prosperous communities living by the east coast of Yorkshire. The region also had its own tool-making industries. Manby notes that: 'Around Flamborough Head and Filey vast quantities of neolithic flint-working debris have been collected on the cliff tops, usually overlooking the gulleys that provide access to the beaches. Massive primary flakes amongst the debris demonstrate that blocks and nodules had been brought up from the beaches for trimming to workable cores.' Durden has described a specialist flint-knapping industrial site that existed at North Dale, about two miles north of Bridlington. Prepared nodules of flint, as well as rough-outs of tools, were imported from the coast at South Landing, three miles to the east of Bridlington, near Flamborough. Flints of a high quality were found in the glacial tills which outcropped in the cliffs there; they could be collected from the beach as they weathered out and fell. The flint was then taken to North Dale, where specialist knappers produced polished discoidal knives, polished flint axes and polished arrowheads.

The nature of science and discovery is such that we know a good deal about some facets of ancient life and landscape, but much less about others. Certain aspects of the material cultures of prehistoric peoples have proved durable, but the classification of flint tools or pot fragments cannot introduce us to the thought patterns, value systems, self-images, emotions or relationships of ancient individuals and communities. Stone tools are virtually indestructible, and decades of painstaking study have created the necessary expertise to work out a chronology linking different types of artefact to different stages of human development. Carbon-14 dating allows us to attain approximate ages for fragments of organic material recovered from excavation, like roots preserved in peat deposits, charcoal from hearths, charred grain, bones and so on. Past environments can be reconstructed wherever ancient pollen grains have been preserved in waterlogged peaty soils, for each pollen grain has its own tough coat and a form which identifies it as coming from an oak, a lime, cereals, weeds of cultivation, and so on. Even the wing-cases of beetles and the shells of particular snails can be excavated and identified in attempts to reconstruct a long-lost environment. For example, the idea that the primeval lowland forest in Yorkshire was dominated by lime trees is supported by the discoveries of the remains of beetles that lived only in lime woods. Modern researchers are able to re-create the environments occupied by the people of different prehistoric periods, though we are quite unable to re-create what it *felt like* to be a member of, say, a Neolithic society.

Whenever pollen has been preserved by wet, acidic soils, one can expect to be able to build a detailed history of the vegetation of a given countryside and to recognise the changes wrought by humans. In the uplands, the replacement of birch, hazel and alder forest by heather could denote Mesolithic clearance; in the lowlands, the later displacement of elm and lime by cereals and weeds of bare ground, like plantain, would suggest land clearance by Neolithic farmers. Much debate concerns the sharp decline in elm pollen which occurs a little after the dawn of farming. Some attribute this to a prehistoric outbreak of Dutch elm disease – and the beetle which carries the spores of the fungus responsible is now known to have been present. However, since lime and ash *also* declined at this time, human clearances provide the most likely explanation. The decline is often regarded as a consequence of the cutting of elm leaves as fodder for livestock at an early stage

in the adoption of farming. On the North York Moors, it appears to have occurred about 3600 BC and in Holderness around 3900 BC; on the best base-rich soils of the Wolds and the Vale of Pickering, the elm could have been removed at still earlier dates.

In contrast, it is much harder to obtain clear evidence of the ancient settlements, their layout and size and the appearance of their dwellings, or to discover the exact methods of farming employed in Neolithic times. This is not surprising, for, apart from anything else, we must remember that a Neolithic settlement will be about twice the age of a prehistoric one of Iron Age vintage: twice the time available for ploughing to obliterate the remains, or for erosion or deposition to remove or bury the traces. If the settlements lay on the slopes, their remains are likely to have been eroded with the slipping soil;

Mesolithic people frequently hunted in areas like Blubberhouses Moor near Harrogate, burning the vegetation to create open hunting ranges. The great surviving expanses of woodland were gradually removed in the Neolithic and Bronze Age periods. If we ignore the geometric patchwork of late eighteenth-century or Napoleonic enclosures in the middle distance, this vista of moor, heath and woodland has a distinctly Neolithic flavour

if they lay in valleys, they would have been buried by thick accumulations of sediment. This point is amply demonstrated by the discovery of a rich hoard of Neolithic relics at the railway works in York, which had been buried in a gradual accumulation of silt some seven metres thick.

Consequently, although it is suspected that the eastern parts of Yorkshire supported a relatively numerous and affluent population in Neolithic times, very little evidence has been discovered to reveal the domestic life of the first generations of Yorkshire farmers. At Beacon Hill, near Flamborough, a few hearths and hollowed areas were found, along with fragments of pottery, but this tells us very little about the actual homes of the people. If we guess that they lived in dispersed hamlets and farmsteads, with timber dwellings and out buildings of a post, thatch and wattle construction, we may well be guessing correctly, but many more fruitful excavations will be needed to establish the facts. A few Neolithic communities in the Dales continued the traditions of cave life: Queen Victoria cave was occupied; Foxholes cave in Clapdale was defended by an outer wall; and a grave was cut into the stalagmite deposits of Dowkerbottom cave, overlooking Wharfedale. One would also like to know much more about the appearance of the countrysides: were there hedged fields, riverside meadows and open upland commons? Were the first farming techniques of the 'slash and burn' type, with the lime-dominated woodland growing on the better soils being cut and burned to release a wood-ash fertiliser, the cleared ground then being farmed to exhaustion and then abandoned to scrub? How much time passed before permanent fields, which could be invigorated by fallow and manuring, were established? These and many other questions are still to be answered, although it would seem likely that, while shifting farming took place in the more marginal environments, permanent fields would soon have been established on the better grounds. At Heslerton, on the northern flanks of the Wolds, excavations have revealed traces of ten shallow troughs or ditches, aligned east–west, and it can be argued that these marked the boundaries of long, narrow fields, with the channels of local streams providing the north–south boundaries. The authors of the report concerned

write: 'This arrangement would provide a series of "fields" covering a 30 m x 200 m area at maximum.'

One can be sure that Yorkshire's varied terrain affected the patterns of early development. On the chalky lands of the Yorkshire Wolds, some valleys of the North York Moors and the drier limestone bordering the Vale of York, conversion to the farming life is likely to have taken place at a very early date. Clearings would have broadened and merged until only isolated stands of the primeval forest of lime, elm, oak, ash and hazel remained on the steeper slopes, with pockets of alder wood marking the heavier, damper valley soils. When a long barrow at Willerby Wold in the Yorkshire Wolds was excavated, the buried land surface was revealed. It emerged that, before the tomb was built, the land existed as a closely cropped pasture, showing that such tombs stood in working countrysides. (The evidence from this and similar tombs confirmed that oxen that were smaller and had shorter horns than the wild aurochs of the forest were kept, as well as pigs not so far removed from the wild boar. Sheep or goats were introduced from the Continent during this period.) Similarly, the fossil soil found beneath the long barrow at Kilham revealed two phases of woodland clearance and cultivation which were separated by a period when the woodland had regenerated. Whether woodland ever colonised the Wolds is debated, though some authorities consider that the region had a countryside mosaic of pasture and light woodland in the Neolithic period. In any event, it seems that the chalk lands of the Wolds were largely cleared of forest by about 3500 BC.

In those countrysides that were less inviting to the farmer, the changes were muted and elements of the old nomadic life endured. Hunting was still important on the open moors and upland grazings of the Pennines and the high ridges of the North York Moors, and there may also have been a younger tradition of driving herds up from the Vale of York and the lower valleys to graze the high summer pastures on the Pennines and moors. In the Pennines, the forest still ran up the slopes to heights of over 365 metres; within these belts of woodland, the clearings stood as islands, often impermanent and probably serving mainly as pastures. The dry

limestone countryside of the Malham area remained wooded until the Neolithic period was well advanced; though some felling took place in Nidderdale around 3300 BC, more serious deforestation did not begin for almost a millennium. Prehistoric farming made far less headway into the waterlogged areas of the wetlands, though such areas were exploited for fish, wildfowl and seasonal grazings, and wooden trackways across marshy ground, similar to the celebrated examples excavated in the Somerset Levels, have been noted. Such tracks appear to have been built across the alder carr at Skipsea Witthow Mere, close to the Holderness coast, where wooden rods and stakes dated to about 3400 BC have been found in the peat. Trackway building required enormous amounts of light timber, implying that managed coppices already existed in the nearby countryside.

Looking back through time, one is always tempted to seek for golden ages: 'Oh, call back yesterday, bid time return' (Shakespeare, *Richard II*, III. 2). For many 'New Age' romantics, a group not known for letting fact get in the way of fantasy, the Neolithic is a lost paradise. Evidence from Yorkshire and other British regions sadly suggests that our modern disposition towards violent territoriality and xenophobia was shared with Neolithic societies. It tells of ancient wars and skirmishes where bodies were pierced by flint-tipped arrows, and it hints at macabre rituals and ghastly displays of heads or skulls. Even so,

Wharfedale, looking towards Kettlewell village. In Neolithic times, the river will have flowed through the lakes and marshes of the valley bottom; above agricultural clearings formed islands in the wildwood which still extended in places to heights of over 365 metres

there must have been much about the life that was genuinely wholesome. Some countrysides had yet to fill; there were still stands of untouched woodland, and the farming operations of the humans had made the countrysides more diverse, creating pasture, meadow, ploughland, fallow and abandoned croplands, as well as walls and, probably, hedgerows. Thus the first phases of human cultivation opened up a spectrum of new ecological niches for colonisation by the plants and animals of more open country. Rabbits, grey squirrels and fallow and sika deer were not to be introduced for thousands of years to come, but beaver, bear, boar, wild horses and wolves still flourished in Yorkshire, salmon glided in the unpolluted rivers, and, without herbicides and pesticides, the pastures, ploughlands and meadows must have been vibrant with bird and insect life and spangled with wild flowers such as one will never see today.

As for the people of these times, in a sense we know more about the dead than the living. It can be argued that societies will adopt the religions that match the needs and concerns of their ways

Part of Nidderdale seen from a vantage point on Brimham Rocks. There appears to have been some clearance of woodland in this dale during the Neolithic period, but wholesale felling did not commence until the Bronze Age

of life. Certainly, the early generations of York-shire farmers seem to have abandoned whatever hunting deities they might have inherited from their predecessors, and the new state of living on the land was proclaimed in massive monuments where the bones of the dead and the wooden 'houses' which often enshrined them were then enveloped by mounds of earth. These earthen long barrows were not a Yorkshire invention, but expressions of a great religion which, in its various forms, had captured the minds of communities living in many parts of western Europe. The first long barrows appeared on the Atlantic margins of continental Europe in about 4400 BC, but by 3000 BC the movement had petered out and new sorts of religious monuments were commanding attention. On the downlands of southern England, the rounded mounds of long barrows, often likened to basking whales, are a characteristic part of the scenery. They are less evident in Yorkshire, but from a significant part of the archaeological heritage of the southern margins of the North York Moors and the northern and western fringes of the Wolds.

Many long barrows were ploughed over and destroyed long ago; some were pillaged and devastated by treasure hunters and ham-fisted antiquarians in the eighteenth and nineteenth centuries; but some have survived to be explored by competent modern archaeologists. Much is still to be discovered about the underlying religion, but the barrows surely reflect the importance of land ownership and fertility to communities whose existence depended utterly on the products of their farmlands. These massive, elongated mounds were generally built in prominent posi-tions, where they could be seen silhouetted on the skyline. With the bones of their most notable ancestors lodged crumbling within such a mound, members of a community announced to trav-ellers, neighbours and potential interlopers that the surrounding lands were theirs and had been so for generations. Perhaps, too, it was believed that the human debris from burials and crema-tions might somehow enrich the soil or make amends for the goodness that farming had taken away.

Modern excavations reveal that long barrows are more complicated constructions than their simple earthen covering mounds might suggest. They also show that Yorkshire was already devel-oping a tradition for 'doing things differently'. Long barrows are common in southern England, usually containing the bones from several or many disarticulated skeletons. In Yorkshire, however, the majority of the tombs were associated with a cremation ritual. If we look at the English distri-bution of surviving long barrows, we find a massive concentration in Wessex, on the scarps embracing the Hampshire basin, with a secondary concen-tration on the Lincolnshire Wolds, extending beyond the Humber estuary into the Yorkshire Wolds and southern flanks of the North York Moors. The largest of the known Yorkshire long barrows, at East Heslerton in the Wolds, has been virtually obliterated by ploughing, but this mound was originally a very impressive structure, some 123 metres long, though only about 9 metres wide at its broadest end, and seems to have been enlarged in stages to accommodate new burials.

One of the most informative examples is the long barrow at Willerby Wold, built to overlook a dry valley from the northern scarp of the Wolds. It was 'dug' in the nineteenth century, but sufficient evidence remained intact to merit a re-excavation in 1958–60. The mound here is about 39 metres long by 11 metres wide, but before it was cast up, using earth dug from two flanking trenches, a 'mortuary enclosure' of a trapezoidal shape with a concave timber façade had been erected. The excavators found that the mound had been built over a 'crematorium trench' which ran from the centre of the façade into the heart of the covering barrow. Corpses were laid in this trench and cov-ered in timber and chalk rubble. Then the barrow mound was thrown up and the cremation trench was ignited. The heat generated during the cre-mation cracked the flints and fused the chalk rubble. Similar in length to the barrow at Willerby Wold, but more than twice as wide, is the long barrow at East Gilling, near Yearsley village, about thirteen miles north of York. It is rather pear-shaped in plan, with the mound surviving to a height of about 2.5 metres. It has revealed evidence of the way in which the sanctity of these mounds survived the passing of the earlier Neolithic religion, for Bronze Age people made a burial in a stone box or cist which they inserted into the southern end of the still-revered mound.

More details about the nature of the ancient rituals have come from the excavation of a long barrow at Kilham, also in the Yorkshire Wolds, where an earlier long barrow seems to have been superseded by the construction of a later one. Here, a trapezium-shaped timber enclosure with three entrance gaps was built and two structures were erected inside it. One, placed in the centre, was a puzzling square arrangement of four upright posts; the other seems to have been a timber mortuary house, a 'house for the dead', where bodies would be left to decompose prior to the internment of their bones. It had the form of a timber-walled passage, its walls revetted by an earthen bank, and it opened on to an entrance to the mortuary enclosure. It covered a 'burial pavement', where five disarticulated and three undisturbed burials had been placed on the pavement of earth and chalk slabs. The other side of the enclosure was approached and entered by an avenue of upright timber posts. A mound of earth filled the mortuary enclosure to the west of the mortuary house, then the enclosure was destroyed by fire, and rubble was heaped over the eastern part of the enclosure containing the mortuary house, to produce a mound of deceptive simplicity which entombed the relics of the rituals.

The traces of tombs are fewer in the Pennines and their forms are rather different. Giant's Grave, at the head of Pen-y-Ghent gill, was an important tomb with stone-built burial chambers beneath an earthen mound of about 15 metres in diameter, while the misleadingly named Druid's Altar, on Malham Moor, was possible a megalithic tomb of the passage-grave type. Barrow burial was probably reserved for the élites of English society, but one less prestigious burial was found in 1936, when the body of a woman, probably aged 35–40, was discovered in a grike or natural crevice in a limestone pavement about two miles north of Horton in Ribblesdale. This was not a traveller who had collapsed on the limestone pavement, of a murder victim, or a body that was casually disposed of, for a valuable polished stone axe – probably of stone quarried at an axe factory in the Langdale Pikes – had been placed

with the dead woman. Stylistically, the axe appeared to date from about 3500 – 3200 BC – about the time that forest was being cleared from the Cumbrian axe-making sites. The Langdale axes played a significant part in the clearance of woodland from the Craven uplands, which came to be quite widely occupied during this period. Gilks and Lord point out that, so far, this is the only later Neolithic burial from a grike, but fifteen examples of burials from caves and rock shelters are known, including the Ebolton, Foxholes and Jubilee caves. It appears that, in the Yorkshire Pennines, caves and rock shelters served an unusual burial cult, with some of them being the abodes both of the living and of the dead.

Tomb fashions changed, and the characteristic long barrows of the Neolithic period were destined to be superseded by the blister-like round barrows of the Bronze Age. However, a number of quite different large, circular barrows were built in Yorkshire during the Neolithic period; along the southern flanks of the Great Wold valley there are four gigantic burial mounds capped with chalk rubble, with another example lying to the north of the valley, on Prior Moor, and a sixth, greatly damaged, barrow lying at Garton Slack. Manby (1988) notes that:

> The excavation of these sites in the 19th Century would have been extremely difficult because of their sheer size and the volume of material to be turned over. The largest barrow, Willy Howe, stands on the floor of the Valley, the great trench cut through it on the orders of Albert, 1st Baron Londesborough (then East Yorkshire's greatest landowner) has the appearance of an abandoned railway cutting.

The best-known and second largest of these is Duggleby Howe, near Duggleby village on the flanks of the Wolds and clearly visible from the B1253. It is an artificial hillock some 64 metres in diameter and 6 metres high, though originally it would have been 3 metres higher. In 1890, the antiquarian, J. R. Mortimer dug a trench into the mound; in the centre was found a great grave-pit, about 3 metres square and almost 3 metres deep, while a shallower grave-pit lay to the east. They contained the crouched skeletons of ten

An aerial view of the great Neolithic tomb of Duggleby Howe, appearing as a dark 'blister' in the centre of the picture, just to the left of the hedge-enclosed medieval field strip. Note the hyphenated curving lines of the causewayed enclosure, though the prominent rectilinear patterns are produced by modern farming. The little circle to the left of the large tomb is probably the ditch of a Bronze Age round barrow. (Derrick Riley)

adult and child burials, with grave-goods which included a pottery bowl, a mace head of antler, a delicate knife of polished flint, a flint axe, blades fashioned from boars' tusks, and long pins of bone. Initially, three burials appear to have been made in some sort of timber chamber or structure, with a second burial phase beginning with the burial of a child, and a third series of adult and child burials following. Contemporary with the inhumations were nine or ten cremation deposits placed at varying depths in the core of the tomb. Long after these burials had been made, people with different beliefs deposited a further 43 cremations into the outer 'skin' of the mound as part of a cremation cemetery which extends for an undetermined distance to the north and west of the tomb. The excavation by Mortimer explored less than half of the enormous tomb, and much still remains unknown. Important new insights have been gained, however, and the aerial archaeologist D. N. Riley wrote: '. . . here the results in 1979 were remarkable, [for they revealed] the crop marks of a great ring approximately 370 m in diameter, which surrounded the mound. The ring was broken into a number of lengths with gaps in between, somewhat resembling the ditch of a large causewayed enclosure'. Causewayed enclosures were the great regional (or tribal?) gathering places of the Neolithic and might have merged the roles of the medieval cathedral and fair. The broken ring of ditches surrounding Duggleby Howe is more puzzling; it has also been compared with the giant henges or religious enclosures of Wiltshire.

Three miles (5 km) to the north-west of Rudston is Willie Howe, a tomb with a diameter of 40 metres and a height of 7 metres. It was trenched twice in the nineteenth century, when a central grave-pit was found, but no burials were recognised. Uncertainties about the date of the tomb remained following a modern excavation, for the original central burial in a deep rock-cut pit was found to have been robbed at an unknown date. Two later burials were found elsewhere in the tomb, one of an adolescent who was placed on a bier of chalk blocks and provided with grave-goods of a ritual beaker and a bronze awl, and one, dated to about 2000 BC, of a Bronze Age man who lay in a tree-trunk coffin and was provided with two flint blades.

Evidence of changing burial rituals and some rather gruesome details were revealed by excavators of a tomb at Whitegrounds, about two miles (3 km) south-west of Malton. This was an early Neolithic tomb of the 'entrance grave' type, dated to about 3700 BC. A passage lined by stone slabs and three metres in length provided the burial chamber; its entrance was sealed with cobbles and the whole was covered by an oval mound. When the grave was excavated, three decapitated burials were found, their heads or skulls placed in a separate pile. The archaeologists also found the skeleton of a woman wearing an amber pendant, a male burial, and a child buried with its pet, an aged fox. Around 3250 BC, a large round barrow surrounded by a kerb of stones was built over the old oval tomb and a grave was dug down into the original burial chamber. In it was buried a man equipped with a splendid polished flint axe. The excavation record raises the question of why the three bodies buried earlier had been decapitated. This rite was occasionally practised in the last stages of pagan times; perhaps people still feared that the spirits might otherwise return to haunt them.

In terms of the political and social geography of East Yorkshire in the Neolithic period, Duggleby Howe seems to have been the great ceremonial centre of the western Wolds, with a complex of monuments at Rudston, near Bridlington, serving the eastern Wolds. If Duggleby was the site of Yorkshire's first great ceremonial focus, others were to emerge. About a millennium after the construction of the first earthen long barrows, a new monument, the henge, appeared. The henge is a circular sacred area which is defined by an earth bank and ditch, the ditch normally lying inside the bank, while the earthworks are breached by one, two or more entrances. Henges are as mysterious as the causewayed enclosures. Since they are sometimes associated with standing stones – as at Stonehenge and Avebury – they are thought to have had religious uses; their defensive capabilities are poor, so they could also have been the venues for secular assemblies and trading. In Yorkshire, it is possible that the great henges which border the Vale of York could have been used as markets for stone-axe traders, or they might have been assembly points where herds were gathered in spring prior to their migration to the lofty Pennine pastures. With a diameter of almost 61 metres, Castle Dykes henge, perched on a col between Bishopdale and Wensleydale, is an impressive example from the uplands.

If one looks at a map of the known British henges – and henges are, almost exclusively, a British phenomenon – the eye is instantly drawn to a great concentration of monuments in the Ripon area of Yorkshire. While Ripon was an important centre for early Christian worship, its location must have had an even greater religious significance in distant Neolithic times. For reasons that are now mysterious, the land between the Ure and the Swale, near the junction of the Dales and the Vale of York was embellished with a complex of great religious monuments. The Thornborough henges near West Tanfield consist of three enormous circular monuments: each with a diameter of about 275 metres, and each defined by banks, once about 3 metres high, which stood within ditches that were 3 metres deep and which were separated from the banks by berms or level areas some 12 metres wide. In 1952, the central circle was excavated and it was found that its banks were of great boulders which originally had an icing of white gypsum crystals, brought here in vast quantities from a deposit a few miles downstream on the Ure. Perhaps the henges were garnished in this way to imitate the

The middle henge of the Thornborough circles – part of a great and neglected Neolithic monument. The two parallel dark lines in the lower part of the photograph are the ditches of an older cursus avenue which was cut when the henge was built.
(Cambridge University Collection)

great monuments of the southern chalk lands, which must have glistened brightly on the downs when their chalk was newly dug? It was also found, from aerial photography, that the central henge was built upon a derelict 'cursus'. This was a gigantic ceremonial avenue, at least a mile in length and bounded on either side by ditches which were about 30 metres apart. The juxtaposition could scarcely have been accidental; when the henges were built, they must have assumed the ritual and ceremonial prestige previously represented by the cursus.

The Thornborough henges stand in line, each with entrances to the north-west and south-east. Though plainly visible in aerial photographs, two of the three have been largely obliterated by ploughing; gravel-digging has bitten into the southern circle, while the northern one is protected by a covering of woodland. The holy ground including and surrounding the monuments is dimpled by early Bronze Age round barrows. One of these was excavated to reveal a coffin which had been hollowed from a tree-trunk and contained a skeleton accompanied by a pottery food vessel and a flint knife. Not far away, at Sutton Moor, are three more great henges, all severely damaged by agriculture, and a scatter of small round barrows. Two of the henges have diameters of around 170 metres, while the third is smaller and seems to be aligned with the Thornborough circles. As has been noted, the ceremonial complex in the vicinity of Ripon and the Ure was not the only focus of its kind in ancient Yorkshire. With its light, base-rich soils that were able to support prolonged human exploitation, the Yorkshire Wolds probably constituted the most prosperous portion of our region in Neolithic times and, after the Duggleby era, a new complex of ritual monuments was developed in the vicinity of Rudston. Three great cursuses spanned the countryside here, but although the terminus of one cursus is visible as a low bank, little else can be seen, except from the air. Nearby, however, in the village churchyard, stands Britain's tallest monolith, an enormous slab of gritstone hauled from a section of coast lying 10 miles (16km) away to the north, at Cayton Bay. It is 7.7 metres tall, 1.8 metres wide and 0.8 metres thick; its removal to Rudston must represent a triumph of ancient engineering. Manby (1988) writes that:

> The movement of such a massive stone block to Rudston, 10 miles (16km) due south of the

outcrop, would have been a considerable engineering operation, the journey lengthened by the need to avoid the wetlands of the Lake Flixton Basin and the gradients of the Wold escarpment and its dales. Movement by water would involve a sea passage around Flamborough Head to reach the mouth of the Gypsey Race in Bridlington Bay; a fairly level route along the floor of the Great Wold Valley would then be possible.

If the four cursus avenues below had become redundant by the time that the stone was erected,

---

The Rudston monolith, erected in a part of old Yorkshire which had probably already been hallowed with the construction of great cursus avenues and tombs, perhaps including the great burial mound of Willie Howe

some of their religious celebrity must have lingered on in the area, for the proximity of the imposing stone to the avenues can scarcely be accidental. The topographical spur which carries the monolith, later church and part of the village must have been a place of great significance and was the focus of three of the cursuses.

More than any other period, the Neolithic, in Yorkshire (and elsewhere) was the time when the framework of the countryside was established. It was the period when colonising farmers learned to discriminate between agricultural localities and to establish the connections between different kinds of places. Lands that were cleared of woodland and put to work by the first farmers nestled in cultivated countrysides which had become old and historic by the time that the first copper- or bronze-smiths arrived to introduce the Ages of metal. The beliefs which sustained the early farmers had evolved too, so that the designs of tombs, temples and other holy places were changed and changed again to meet the spiritual needs of the communities. In the process of catering to the demands of belief and ritual, societies gained a command of engineering skills which would not discredit members of a pioneer regiment. At the same time, the establishment of communities which were organised, disciplined and could be compelled to engage in operations like the building of Duggleby or the transport of the Rudston slab indicate the stretching of the social hierarchy and the emergence of leaders of high status. By the close of the Neolithic, large parts of Yorkshire were under the plough or grazed as pasture; some woods and marshes retained a primeval aspect, though all had timber, browse, fruits, fur and flesh to contribute to the tribal economies. There were no secret places left, and already many of the more fragile environments had more than once become degraded and abandoned following phases of agricultural misuse.

The River Ure near the Thornborough circles. With a name which denotes the sanctity of the river in Iron Age times, the Ure was probably sacred to the Neolithic people, who created a great complex of ritual monuments on the adjacent plains

REFERENCES

T. C. M. Brewster, 'Five Yorkshire barrows' *Current Archaeology*, 94 (1984), pp. 327–33.

P. C. Buckland and K. Edward, 'The longevity of pastoral episodes of clearance activity in pollen: the role of post-occupation grazing', *Journal of Biogeography*, 11 (1984), pp. 243–9.

T. Durden, 'Later Neolithic flintwork: Yorkshire Wolds', *Proceedings of the Prehistoric Society*, 61 (1995), pp. 409–32.

D. P. Dymond, 'Ritual monuments at Rudston, E. Yorkshire', *Proceedings of the Prehistoric Society*, 32 (1966), pp. 86–95.

G. G. Garbett, 'The elm decline: the depletion of a resource', *New Phytology*, 88 (1981), pp. 573–85.

D. D. Gilbertson, 'Early Neolithic utilisation and management of alder carr at Skipsea Witthow Mere, Holderness' *Yorkshire Archaeological Journal*, 56 (1984), pp. 17–22.

J. A. Gilks and T. C. Lord, 'A late Neolithic crevice burial from Selside, Ribblesdale, North Yorkshire' *Yorkshire Archaeological Journal*, 57 (1985), pp. 1–5.

J. B. Innes and I. G. Simmons, 'Disturbance and diversity: floristic changes associated with pre-elm decline woodland recession in north-east Yorkshire', in M. Jones (ed.), *Archaeology and the Flora of the British Isles*, (Oxford University Committee for Archaeology: 1988), pp. 7–20.

T. G. Manby, 'The excavations of the Willerby Wold long barrow; East Yorkshire' *Proceedings of the Prehistoric Society*, 29 (1963), pp. 173–205.

T. G. Manby, 'Neolithic occupation sites on the Yorkshire Wolds' *Yorkshire Archaeological Journal*, 47 (1975), pp. 23–59.

T. G. Manby, 'Excavation of the Kilham long barrow, East Riding of Yorkshire', *Proceedings of the Prehistoric Society*, 42 (1976), pp. 111–59.

T. G. Manby, 'The Neolithic period in Eastern Yorkshire' in T. G. Manby (ed.), *Archaeology in Eastern Yorkshire: Essays in Honour of T. C. M. Brewster FSA* (Sheffield: 1988), pp. 35–88.

D. Powlesland, with C. Haughton and J. Hanson, 'Excavations at Heslerton, North Yorkshire 1978–82', *Archaeological Journal*, 143 (1986), pp. 53–173.

J. Quartermaine and P. Claris, 'The Langdale axe factories', *Current Archaeology*, 102 (1986), pp. 212–13.

D. N. Riley, 'Recent air photographs of Duggleby Howe and Ferrybridge Henge' *Yorkshire Archaeological Journal*, 52 (1980), pp. 174–8.

D. N. Riley, 'Air survey of Neolithic sites on the Yorkshire Wolds', in T. G. Manby (ed.), *Archaeology in Eastern Yorkshire: Essays in Honour of T. C. M. Brewster FSA* (Sheffield: 1988), pp. 85–93.

B. E. Vyner, 'The excavation of a Neolithic cairn at Street House, Loftus, Cleveland', *Proceedings of the Prehistoric Society*, 50 (1984), pp. 151–195.

P. R. Wilson, 'Excavation at East Gilling long barrow', *Yorkshire Archaeological Journal*, 60 (1988), pp. 1–3.

CHAPTER FOUR

Yorkshire in the Bronze Age

AS THE NEOLITHIC PERIOD DREW TO ITS CLOSE, Yorkshire must, in many ways, have been a very attractive place: a mosaic of older and newer farmlands, forest, moor and marsh. By the close of the Bronze Age, however it has become a land that was quite heavily populated, where boundaries were jealously guarded and where land hunger and insecurity underlay the building of defence works. These tensions may also help to explain the strengthening of the distinctly stratified society, which is indicated both by the surviving monuments and by the rich grave-goods which accompanied the burials of those of the privileged caste. Far more people lived in Bronze Age Yorkshire than ever found a resting place beneath a massive barrow mound, and those whose relics are exposed by excavations must have been members of the leading families of their day. In the earlier part of the Bronze Age, a short-age of land in the most attractive agricultural areas resulted in considerable upland settlement and land clearance; in the latter part of the era, a distinct deterioration in the climate, combined with the adverse effects of deforestation and agri-culture, caused a wholesale migration from the uplands – which further increased competition and insecurity in the lowlands.

'Bronze Age' is not a very satisfactory label for this slab of time. It was only very gradually that bronze (which was introduced just before 2000 BC) partially supplanted the more traditional tool-making materials – and, in any event, the use of bronze was preceded by a long phase of copper-working. Around 2750 BC, just a little before the arrival of copper-working, Britain experienced a revolutionary invasion. Whether this was an invasion of *people* or of continental *ideas* is still hotly debated, but the 'Beaker period'

marked a return of Britain to mainstream European thinking about belief and material cul-ture, and it is seen as a distinctive stepping stone between the Neolithic period and the Bronze Age proper. The name derives from the new practice of providing the dead with a finely made pottery drinking vessel or beaker; also, according to the new rites, corpses tended frequently to be buried singly, in a crouched position, beneath a round barrow or in a 'box' or cist of stone slabs. In the North York Moors, traces of more than 10,000 round barrows have been recorded, though some of these will be no more than cairns of rubble built to hold the stones that were gathered as land was cleared for farming. The new burial practices were not altogether standardised, but they do mark a break with the old habits in ritual engineering represented by the long barrows and causewayed enclosures, although, in some other parts of Britain, established traditions rep-resented by the henge and the stone circle were developed to produce some absolutely spectacular monuments.

Attempts to visualise the Yorkshire countryside of the Bronze Age are thwarted by the same sorts of problems of human destruction, erosion and silting that blur visions of the Neolithic. Even so, the evidence is a little richer and it shows us that different parts of Yorkshire were exploited in different ways. In the North York Moors, an area described by archaeologist Frank Elgee in 1930 as 'an open-air museum of the Bronze Age', the sporadic and temporary woodland clearances of the earlier periods were superseded in many places by more mature farming countrysides, complete with networks of fields. A recent careful scrutiny of the evidence enabled Spratt to recon-struct a typical 'estate' of the period. The main

farming settlements would lie in the bottoms of the valleys, with the walled fields of their plough-lands and pastures filling the countryside around them. Each community would also enjoy its rights to an expanse of higher ground on the adjacent flanks of the uplands. This was land that was occupied in the summer season, when cattle and sheep were pastured, and where a little arable farming would take place. The most striking monuments to the vanished Bronze Age lifestyle are the 'cairnfields', areas of ground that are still strewn with small cairns or mounds of gathered stone which are about three metres in diameter. Some cairnfields have just a few cairns, while others, like Iron Howe on Cow Ridge in the north-western part of the North York Moors, contain hundreds. It was once thought that the cairns were tombs; although a few of them may mark burials, the great majority seem to be repositories for the stones that were gathered while preparing the ground for cultivation. In the neighbouring areas, one may also recognise the tumbled and overgrown relics of Bronze Age field walls – long continuous walls running roughly parallel, with shorter walls set between them to partition the land into small rectangular fields. On the higher ground above the cairnfields and enclosed farmland, the woods and moors continued their traditional role as hunting grounds and grazings, with pastures established in the clearings. Most interestingly, the high ridges and watersheds were

The North York Moors consist of heather- and bracken-covered plateaux dissected by fertile valleys. During the Bronze Age, haymeadows may also have linked the valley floor of Danby Dale, but there would have been many more farmsteads and hamlets and a more intensive exploitation of the countryside

punctuated by round barrows. Built of gathered stones, these large cairns are 7–35 metres in diameter and, standing in long lines which span the ridge tops, they seem to demarcate the boundaries of the Bronze Age estates. These barrows were tombs as well as boundary markers, but the devastation caused by the Victorian anti-quaries, who pillaged far more than was ever recorded, has left us with little evidence of their contents. More recent archaeological research on the fossilised pollen evidence from the ground on which some of these barrows were built suggests that they were raised in clearings cut into a light woodland cover.

This reconstruction is based on the evidence from the sandstone moors and valleys of the North York Moors, while the limestone hills of the region – the Hambleton Hills which form the western rampart of the moors, and the Tabular Hills overlooking the Vale of Pickering – seem to have been very well peopled at this time. Like the villages of the historical period, the main Bronze Age settlements were probably aligned along the

Chains of round barrows lined the watersheds and delimited the Bronze Age territories of the North York Moors. This crumbling example stands with others on the summit of Beacon Hill–hence the beacon structure. Ruined 'howes' like this still punctuate the moors

junction of the limestone hills and the vale soils, and a cluster of round barrows around Pickering itself suggests the popularity of this area in Bronze Age times. Further south, the Yorkshire Wolds continued to be a particularly favoured region, as evidenced by its imposing assemblage of round barrows. An organisation of the country-side similar to that described in the North York Moors is likely to have existed here. The hills and valleys of the eastern Wolds were probably sheep pasture, with the Bronze Age round barrows tending to be concentrated in this latter area. The fact that later Bronze Age estates on the Wolds could be delineated by lengthy banks and ditches argues that the countrysides were largely cleared of woodland. Flint scrapers, suitable for scraping flesh from hides, are very frequently found in the Wolds, arguing for an economy focused on live-stock rather than on cereal cultivation.

The great Vales and the lowlands of Holderness are more reluctant to yield their secrets. Thick accumulations of silt blanket most of the traces of ancient settlement here, though building or pipe-laying operations occasionally produce the relics of prehistoric life and, under favourable circumstances, aerial photography reveals the rec-tangular ditches which enclosed some farmsteads.

Much of the lowest ground would have been poorly drained and marshy, so that settlements and farms would have congregated on the drier sites. Damper fields could have served as summer cattle pastures, while the rivers and marshes would have continued to yield generous harvests of fish and wildfowl.

New woodland clearings were made in the Pennines, though they were often of a small and temporary nature. There seems to have been considerable variation in the upland countrysides, though the emphasis was generally on hunting and grazing rather than on the cultivation of grain. In some places, attempts to extend the grassland by clearing the forest may have been thwarted by the establishment of heather moors rather than the hoped-for green pasture. In other places, grass grew beneath a light covering of hazel; the lower slopes are likely to have been pasture, but, in many places, a zone of deciduous woodland will have clothed the slopes between these fields and clearings and the high moors. Most of the settlement was probably in the foothills and valleys, where fishing and fowling could supplement farm production. Although the population supported was more numerous than before, summer migrations to exploit the game and grazings of the higher grounds continued.

Religion played a prominent part in the lives of the Bronze Age communities. Vyner has recently suggested that, on the Cleveland Hills, areas for religious rituals were defined by the digging of cross-ridge boundaries which cut off upland spurs or promontories, usually at their narrowest points and typically by an earthwork consisting of a single bank and ditch. To the south of the River Esk, a series of spurs, as at Ingleby Greenhow, Castleton, Danby and Glaisdale, are cut in this way. Far better-known are the numerous round barrows of eastern Yorkshire. About 200 Bronze Age barrows have been excavated on the North York Moors, most of them by Victorian antiquaries, whose methods were crude. Spratt (1989) writes that:

Of the 200 barrows, some four-fifths contained Bronze Age pottery and three-fifths had cremated human bones, but, usually, very little else. Not infrequently the barrows contain later cremations inserted into their structures as well as the early ones at their very centre . . . Very occasionally on the North York Moors, however, we find barrows, placed at panoramic viewpoints, which contain more elaborate articles, sometimes of bronze; these barrows seem to be memorials of high status families. There is one at Swarth Howe, standing on a conspicuous position opposite the car park 5½km (3½miles) along the Guisborough road inland from Whitby.

People who had adopted the Beaker package of beliefs and rituals generally tended to favour solitary burials, but in Yorkshire more sociable – or, perhaps, thrifty – customs often prevailed. Beneath many of the excavated round barrows here archaeologists have found large pit-graves, in which several successive internments were made before the dome-shaped covering mound was at last cast up. Between 1964 and 1975, a special excavation project explored some 34 round barrows in the east of Yorkshire, and the results showed just how varied the burial customs could be. The example explored at Hutton Buscel, on moors five miles (8km) to the west of Scarborough, was a splendid monument ringed by a kerb of stones, some grooved through use in sharpening axes, some decorated with cup-like hollows and pecked geometrical motifs. The barrow was cast up over the burial of an 8- or 9-year-old child, who was provided with seven flint blades. Instead of the ritual beaker that one might have expected to find with the burial, a miniature 'food vessel' had been provided; another

was placed just inside the stone kerb. Yorkshire food vessels were more robust and more bowl or vase-shaped than beakers and might have contained a ritual offering of food for the dead (the beakers generally being assumed to have contained an alcoholic drink). They were particularly popular in the Yorkshire Wolds, but they are also quite common in the Peak District, north-east England and eastern Scotland. It was thought that the food-vessel cult succeeded that of the beaker, but in fact some food vessels were contemporaneous with beakers.

In Bronze Age society, the differences in the social and economic roles of the sexes were recognised after death. Pierpoint writes that:

Men at burial were particularly associated with plano-convex knives and other flint tools and arrowheads, stone battle axes, and daggers of flint or bronze; females were associated primarily with a restricted range of beads, bone pins, and copper and bone awls. The same kinds of arrowheads as those buried with males in barrows are also found as the main item of bronze age lithic material in many upland areas of the wolds and moors: their frequency in these areas suggests that these were hunting zones, with the implication that one area of the male role in bronze age society in Yorkshire was that of hunter. Certainly simple distinctions between male and female roles in the subsistence system are very common in primative agricultural societies.

Another Yorkshire rite was that of the 'shaft grave'. When a barrow at Octon Wold near Bridlington was excavated, it was found to cover two such graves. The skeleton of a woman lay at the bottom of a shaft, 2.1 metres deep, while another shaft, about 1.8 metres deep, contained the remains of five people and had been opened three times before the final sealing. The last burial was that of a man placed in a flexed position in a stone bier, with two of the skulls from the earlier burials behind his pelvis and limb bones placed on his chest. He died around 2200 BC. Three other graves, not in shafts, were also found within the barrow. Yet another Yorkshire rite is represented by the 'cremation furnace barrow'. Two such barrows were excavated on Sawdon Moor, 7 miles (11 km) west of Scarborough. One example contained two cremation furnaces: shallow pits where bodies were cremated. A date of about 1800 BC was obtained from cremated materials. After the initial barrow of sand was built over the cremation furnaces, the monument was enlarged and surrounded by a circular kerb of stones of 16 metres in diameter. As at several other excavated Bronze Age barrows in Yorkshire, there were clear traces that the site covered by the tomb had previously been occupied by Neolithic people, the evidence here consisting of a hearth, pottery, flints and a flint axe. As a final example, one can quote the barrow at Loose Howe, on Danby High Moor in the North York Moors. Here a mound of 18 metres across and 2.1 metres high covered the burial of a warrior armed with a bronze dagger and dressed in linen garments and leather shoes. A dugout canoe about 2.7 metres in length was used as a coffin, a second canoe covered the coffin and a third was placed alongside. Within the coffin, the body lay on a bed of reeds, rushes and straw, the head resting on a straw pillow. Within the same mound there was a second burial of cremated bones contained in an urn with a stone battleaxe and a bronze dagger as grave-goods.

Yorkshire is rich in Bronze Age tombs, even though many have been destroyed. Although these mounds seem superficially similar, the expertly excavated examples reveal that a host of different rituals were practised during the currency of the round barrow. People might be buried in shafts or pits, in tree-trunk coffins or stone cists; alternatively, they might be cremated. Beakers or food vessels, along with simple possessions, could accompany the dead, or there might be no grave-goods at all. Burials could be solitary, but several simultaneous or successive interments were often made. All this makes it very difficult to imagine that there was a single, codified Bronze Age religion in Yorkshire. The region had already demonstrated its cultural tendency to do things rather differently, and it may be that dynasties or local cults adopted the Beaker package of beliefs and rituals and then developed their own variations on the themes until the original features of the religion were almost unrecognisable.

The era of the round barrow overlapped with that of another expression of Bronze Age religion: the stone circle. Given the numerous round barrows and the imposing collection of monoliths that Yorkshire can boast, it is maybe surprising that there are no really impressive stone circles – though the great henges built between the Ure and the Swale were probably erected by people influenced by the Beaker culture in the transition from the Neolithic to the Bronze Age. These henges were probably built in the period 2700–1600 BC. There are a number of small circles, but they should sometimes be more properly regarded as the kerbs which surrounded severely eroded barrows. One of the most charming, but also most dubious, 'stone circles' is right beside the Wharfe at Yockenthwaite. Twenty stones are set edge to edge to form a ring with a diameter of 7.6 metres; the appearance is that of a tomb kerb rather than a true stone circle. Several small stone rings are found on the North York Moors, all of them damaged and most of them really tomb kerbs rather than circles. At Danby Rigg, just one tall standing stone, 1.5 metres high, remains from the circle; the nearby barrows and cairnfield, with about 300 clearance cairns, are more impressive. Twenty-four small stones form a circle with a diameter of 10 metres at Standing Stone Rigg, on the moors 6 miles (10 km) to the north-west of Scarborough; the remains of a burial chamber at the centre of the circle suggest that this monument was the retaining wall of a barrow. This interpretation also applies to the Bride Stones on Bilsdale East Moor, where 40 small stones define a circle with a diameter of 12.2 metres. A more unusual monument is the Druid's Altar on Malham Moor; it is thought by some to be a far-flung member of a community of stone circles found in Scotland and known, picturesquely, as 'four-posters', though only three of the original four uprights still stand. Others have interpreted it as an older megalithic tomb of the passage-grave type.

What Yorkshire lacked in stone circles it made up for in standing stones. The great monolith at Rudston has already been noted; on the outskirts of Boroughbridge stand the Devil's Arrows, three enormous pillars of gritstone. The monument is described by the prehistorian Aubrey Burl as 'one

The small stone circle at Yockenthwaite is more likely to represent the kerb of stones encircling a Bronze Age tomb

of the most astonishing megalithic settings of western Europe'. The tallest stone stands some 6.9 metres above the ground, and all have stood upright for so long that the grit at their tips has been fluted by the dripping rainwater. The most likely source for their gritstone was at Plumpton Rocks near Knaresborough, and the transport of such gigantic slabs over a distance of at least nine miles (13 km) is another remarkable feat of pre-historic engineering. Some special qualities must have been associated with the Knaresborough stone, for perfectly good sandstone was available at the Boroughbridge site. Burl noted that the lightest of the stones weighs over 25 tons and he estimated that 135 labourers would have been needed to haul the heaviest stone upright. He thought that the transport of the stones must have taken place over several years. The stones stand in a line 174 metres long and orientated NNW–SSE; how frustrating it must be for the advocates of prehistoric geometry and astronomy that the Arrows are not aligned exactly! There was originally a fourth stone; Leland, the anti-quary, visited the stones around 1535–40 and mentioned that they were four in number and,

three decades later, William Camden mentioned that one of the stones had been pulled down by treasure hunters. It was probably dragged to Boroughbridge later to form part of the founda-tions of a bridge built over the River Tuff in the 1620s. Burl wrote: 'That some of the tallest stand-ing stones in the British Isles, at Rudston and the Devilís Arrows, should be found in Yorkshire, a region not [otherwise!] noted for its megaliths, on sites many miles from the source of the stones, suggests that there were compelling reasons for their erection.' He hypothesised a date for the erection of the stones of 2200–2000 BC, which would make them contemporaneous with the sarsen stone circle at Stonehenge. The nature of the relationship of the Devil's Arrows to the chain of – possibly older – sacred sites associated with the henges at Thornborough, Nunwick, Cana and Hutton Moor, which extends northwards from the Arrows for 11 miles (16 km) is not known.

Two of the Devil's Arrows trio of massive monoliths (the third lies behind the photographer and is obscured by trees)

So far as quantities of Bronze Age monuments are concerned, the only other area in Yorkshire that can rival the assemblages of the North York Moors is Ilkley Moor. Here there are eight stone circles, mostly barrow kerbs really, the most notable being the Skirtful of Stones and the Horncliffe, Grubstones and Twelve Apostles circles. Dotted around the moor are boulders decorated with crude carvings: cup and ring marks, ladder patterns, and, on one boulder, swastikas. The meaning of these symbols, which are also found on boulders, rock outcrops and religious monuments in some other parts of upland Britain, remains a mystery. No less mysterious are the 'segmented embanked pit-alignments' recently recognised in the North York Moors. Pit alignments are normally chains of pits which cross the countryside like a dotted line, generally assumed to mark prehistoric boundaries. However, these earthworks consist of two or three pairs of pits, with the spoil from the pits spread into parallel enclosing banks. Nothing comparable is known

elsewhere in England or Scotland, but Lofthouse reports that these earthworks have been noted on the moors at Middle Rigg, Easington High Moor, Ugthorpe Moor and Danby Rigg. They are made up of several short sections, generally orientated north-west to south-east, and seem to have tangential alignments on Bronze Age burial mounds.

By around 1600 BC, the fascination with grandiose tombs and standing stones was exhausted. In some places, 'graveyards' were created, stocked with cremations contained in urns, but in most places it is impossible to identify any late Bronze Age or Iron Age funeral rite (apart from that associated with the square Iron Age barrows of eastern Yorkshire, a very special class of monuments). Probably, religion had drifted – or drifted further – in the direction of the worship of water spirits. According to the evidence discovered, corpses, sacrificial victims and votive offerings of costly weapons were cast into rivers, lakes and marshes.

The concept of the Bronze Age as a distinctive period has come under heavy criticism, and it might be more realistic to divide the era into two halves. The first half saw an intensification and an extension of the farming activities of the previous age and a continuing interest in the creation of

spectacular religious monuments. The second half of the Bronze Age continues easily into the Iron Age and is marked by the abandonment of farming in some places, population growth in others, rising levels of insecurity and a lack of interest in recognisable religious constructions. Many of the changes can be linked to a deterioration of the climate. As early as 1500 BC, a trend towards wetter, cooler conditions was leading to the abandonment of farming on the Pennine moors. Where generations of farmers had cleared the woodland and exhausted some lands, so the rains leached out the goodness from the soil and the ground soon became covered in a blanket of sour, waterlogged peat. Similarly, on the gritstone uplands of the North York Moors, old farmlands degenerated into rough grazings, and, as peat and heather spread across the high ground, hunting camps which might have been visited for 5000 years were finally abandoned. In the lowland valleys, meanwhile, a rising sea level caused riverside marshes to broaden out across the flood plains.

Whether these changes stimulated a special interest in navigation, we do not know, but the region has yielded some particularly important relics. A number of prehistoric vessels have been discovered around the mouth and flanks of the Humber estuary. Three examples were found at North Ferriby, just to the west of Hull, which date to around 1500 BC. We have seen that the people of these times were capable of building dugout canoes, but a reconstruction of one of the North Ferriby boats reveals a much more sophisticated craft, with the planks joined and stitched together around bow-shaped formers to create a seaworthy craft. Other tantalising evidence from the Humber is provided by the Roos Carr model, boat which could also belong to the Bronze Age and represents

A spectacular expanse of limestone pavement emerging through melting snow above Malham Cove. Deforestation, followed by overgrazing in the prehistoric period, would have encourage soil erosion and the exposure of such pavements

in miniature a craft manned by warriors with round shields.

Although the population of Yorkshire must have continued to rise through this period, lowland flooding and the advance of peat across the upland moors reduced the amount of land available for farming. As communities abandoned their traditional lands, so the competition for good farmland intensified. The development of efficient bronze swords added to the horrors of warfare, and it was at this time, if not before, that society became markedly stratified and dominated by well-armed and gorgeously attired chieftains and warriors. By 1000 BC, England may have been as heavily populated as it was in early medieval times, but the continuing environmental problems exacerbated the tensions. The growing proficiency of the bronze-smiths allowed members of the warrior class to acquire formidable weapons – like the bronze rapier dating from the twelfth or eleventh century BC which was discovered in 1992 during the building of a river-gauging station on the Swale at Catterick Bridge. More traditional, but no less deadly, were the stone battleaxes, many of the best ones in Yorkshire being shaped from dolerite quarried from igneous dykes and sills in Northumberland and Durham, like the sill known as the Whin Sill, which carries Hadrian's Wall.

The troubles of the time are reflected in surviving evidence from the North York Moors: on the Hambleton Hills, the territorial boundaries which had been marked by barrows were now made more prominent by the digging of several miles of banks and ditches; in the Yorkshire Wolds, the building of round barrows petered out around 1600 BC, and enormous efforts were subsequently devoted to the construction of linear boundaries or 'entrenchments' around the estates. These earthworks divided the countryside into large enclosures; although it is not known for certain whether they were constructed as boundary markers or trackways, they indicate a fully exploited landscape, with no unclaimed spaces. They seem to have appeared after the middle of the Bronze Age and to have been extended and elaborated during the Iron Age. All these earthworks must have had a territorial function, to warn neighbours and vagrants of the perils of trespass and land-grabbing, but more practical defence works were also needed. Farmsteads were protected by strong palisades and ditches, and a number of citadels are known to have been established in Yorkshire at the close of the Bronze Age.

Eston Nab hill-fort, on the northern fringe of the North York Moors overlooking the Tees, is a heavily fortified Iron Age enclosure which is about 91 metres wide by more than 183 metres long, but it began as a much smaller Bronze Age palisaded defence work. Almondbury hill-fort near Huddersfield has a very complex history, although most of the more prominent earthworks belong to the Norman fortifications. It began in the eighth century BC as a defensive enclosure with a single rampart about four metres wide. Then the fort seems to have been allowed to decay, and several circular dwellings were built on the site before a second phase of fortification was launched. A new rampart was built in roughly the same position as the old one, and then a second ring of ramparts doubled the defended area and secured the whole plateau area. Subsequently, around 600 BC, a third rampart was built, which encircled parts of the slope below. At about the end of the fifth century BC, the fort was destroyed by burning; no new ramparts were raised until the Norman works in the medieval period.

Grimthorpe, on the eastern edge of the Yorkshire Wolds, and Staple Howe, a steep knoll on the northern fringe of the Wolds, were other early hill-forts. Grimthorpe offers nothing to the casual visitor, but excavations date the fort to about 800 BC. At Staple Howe, where fortification began later, a simple palisade was built to enclose an oval area measuring about 59 metres by 26.5 metres which contained an oval house. The dwelling was almost 9 metres in length, with stone walls and a gabled roof, and contained a hearth and a clay oven. Subsequently, a rampart was built lower down the hill to enlarge the defended area. The oval house was replaced by two circular ones, with strong timber walls and conical roofs of thatch, and a raised granary or a watch-tower was built in the centre of the enclosure. Staple Howe could have been the defended home of one of the more affluent and influential farming families. They hunted and fished, kept cattle, goats, sheep and pigs, and grew cereal crops. The site was identified and excavated by the late T. C. M. Brewster, following a programme of field walking; Powlesland writes: 'The degree to which almost the whole population of a small Yorkshire village became involved in the Staple Howe excavation is a tribute to Brewster's skill and enthusiasm which has generated a local understanding and awareness of archaeology which remains exceptional today'. Brewster went on to discover a similar site at Devil's Hill, 1⅓ miles (2km) to the east.

Valuable information about the landscape in which these defended, aristocratic settlements existed has resulted from the Heslerton survey. It emerged that land divisions seem to have been established in the Bronze Age by the digging of long chains of pit alignments, and that the recutting of these features could have produced the entrenchments which seem to define estate boundaries. Powlesland writes that:

> Given the topographical relationship of the two excavated palisaded enclosures, Staple Howe and Devil's Hill, to the identified pit alignments, it is difficult to argue that these are not related elements in a well organised landscape. The physical labour involved in the creation of these boundary features implies both a considerable population and a highly organised society regardless of what function it assigned to the boundary features themselves.

An undefended Bronze Age settlement of strung-out dwellings developed along a pit alignment, running from west to east. Powlesland goes on to suggest that the estate centres may perhaps be related to settlement by early Celtic colonists of the continental Hallstatt culture: 'The Vale of Pickering may have formed a focus for Hallstatt settlement, perhaps by an elite who settled on the commanding sites at Staple Howe, Devil's Hill, and others yet to be located. Associated with these high status enclosed settlements, open settlements of relatively low status appear to have grown up along the line of an east–west aligned trackway and a major boundary on the flat lands of the Vale . . .'. There is evidence that some of the boundaries that were established to mark out the territorial spheres in eastern Yorkshire during the latter phases of the Bronze Age and the early Iron Age endured into the medieval period and may still survive as parish boundaries.

Although some of the early hill-forts exploited the defensive possibilities of the terrain, others seem to have guarded lowland farming areas. For want of a better name, these are referred to as 'lowland hill-forts'. Excavations near Thwing, on a spur in the north-eastern part of the Yorkshire Wolds, explored a 'mini-hill-fort' of the late Bronze Age, built around 1000 BC. A circular defensive ditch some 3 metres deep was cut into the chalk to defend an area of about 108 metres in diameter. A rampart was built inside the ditch, and an inner ditch was dug within the rampart to a depth of 1.8 metres. At the centre of the defended area contained a circular structure with a diameter of around 27 metres. Whether this existed as an open circular arrangement of posts, or whether these posts could have carried an enormous roof, is uncertain. With its formidable defences, the Thwing fort could easily have served as a local defensive nucleus, but it could also have been used as a centre for religious celebrations or secular gatherings. A great deal of domestic material was found: querns and rubbing stones for grinding grain, loom weights, spindle whorls, bronze pins, awls, rivets and rings. In addition, the excavation yielded large quantities of animal remains, showing that the local community kept cattle, sheep, pigs and horses, and hunted deer. A careful examination of the animal remains suggested that the cattle were kept mainly for milk, with the surplus bulls and unproductive females being slaughtered for meat; the sheep were kept primarily for wool. It also appears that, before the fort was built, the villagers here existed as a self-sufficient community, killing animals as and when meat was required to meet their domestic needs. Later, however, the fort became a centre for ritual or the payment of tribute, with animals being brought here for slaughter in considerable numbers. Mounteney wrote that:

> The contrast between the material from the pre-rampart and terminal deposits could reasonably be interpreted as the result of economic change at the site, from a more or less self sufficient village slaughtering animals as and when required for domestic needs, to a more complex economy in which Thwing became the centre of some kind of regional redistribution or tribute system, with groups of selected animals being brought to the site for slaughter in quite substantial numbers on separate occasions – although whether sporadically or frequently we cannot tell.

If these interpretations are correct, then the Thwing fort would seem to epitomise the general situation in the second half of the Bronze Age, with the parallel eruption of strongholds and commanding leaders. Perhaps the land hunger and social tensions of the age fostered the emergence of warrior chiefs whose citadels became the focuses for defence and the centres to which the people of the surrounding 'estate' were obliged to render tribute. This reminds one that, as in other times, that the power and the glory of the Bronze Age were founded on peasant toil. If the framework of the countryside was established in the Neolithic period, the Bronze Age was a time of filling-in and consolidation: between the abandoned moors, with their wooded fringes, and the marsh and carr of the valley bottoms, the landscape was now cleared and hard at work, with only patches of useful woodland standing as remote islands amongst the ploughlands and pastures.

REFERENCES

T. C. M. Brewster, *The Excavation of Staple Howe* (Scarborough: 1963).

T. C. M. Brewster, 'The Devil's Hill' *Current Archaeology*, 76 (1981), pp. 140–1.

T. C. M. Brewster, 'Five Yorkshire barrows', *Current Archaeology*, 94 (1984), pp. 327–33.

C. Burgess, 'A bronze rapier from Catterick Bridge', *Yorkshire Archaeological Journal* 67 (1995), pp. 1–5.

A. Burl, 'The Devil's Arrows', *Yorkshire Archaeological Journal*, 63 (1991), pp. 1–24.

D. Coombs and T. G. Manby, 'The excavation of two Bronze Age round barrows on Irton Moor, Yorkshire, 1973', *Yorkshire Archaeological Journal*, 66 (1994), pp. 21–50.

F. Elgee, *Early Man in North-East Yorkshire* (Gloucester: 1930).

C. A. Lofthouse, 'Segmented embanked pit-alignments in the North York Moors: a survey by the Royal Commission on Historical Monuments of England', *Proceedings of the Prehistoric Society*, 59 (1993), pp. 383–92.

T. G. Manby, 'The Yorkshire Dykes', *Current Archaeology*, 67 (1979), p. 233.

T. G. Manby, 'Bronze Age settlement in eastern Yorkshire', in J. Barrett and R. Bradley (eds), *The British Later Bronze Age*, British Archaeological Report 83 (1980), pp. 307–64.

P. Mayes, M. Atherden, K. Manchester and T. G. Manby, 'A Beaker burial at West Tanfield, North Yorkshire', *Yorkshire Archaeological Journal*, 58 (1986), pp. 1–4.

G. Mounteney, 'Faunal attrition and subsistence reconstruction at Thwing', in G. Barker (ed.), *Prehistoric Communities in Northern England* (Sheffield: 1981), pp. 73–86.

H. Mytum, 'A battle axe from Appleton Roebuck', *Yorkshire Archaeological Journal*, 60 (1988), pp. 175–6.

S. J. Pierpoint, 'Land, settlement and society in the Yorkshire Bronze Age' in G. Barker (ed.), *Prehistoric Communities in Northern England* (Sheffield: 1981), pp. 41–56.

D. J. Powlesland, 'Staple Howe in its landscape', in T. G. Manby (ed.), *Archaeology in Eastern Yorkshire Essays in Honour of T. C. M. Brewster FSA* (Sheffield: 1988), pp. 101–7.

D. A. Spratt, 'Prehistoric boundaries on the North Yorkshire Moors' in G. Barker (ed.), *Prehistoric Communities in Northern England* (Sheffield: 1981), pp. 87–104.

D. A. Spratt, 'The prehistoric remains', in D. A. Spratt and B. J. D. Harrison (eds), *The North York Moors: Landscape Heritage* (Newton Abbot: 1989), pp. 28–44.

W. J. Varley, *Castle Hill, Almondbury: A Brief Guide* (Huddersfield: 1973).

B. E. Vyner, 'The territory of ritual: cross-ridge boundaries and the prehistoric landscape of the Cleveland Hills, northeast England' *Antiquity*, 68 (1994), pp. 27–8.

D. R. Wilson, 'Pit-alignment distribution and function', in H. C. Bowen and P. J. Fowler (eds), *Early Land Allotment*, British Archaeological Report 48 (1978), pp. 3–6.

CHAPTER FIVE

The Brigantes and Parisi

THERE IS NO SHARP DIVIDE BETWEEN the Bronze and Iron Ages in Yorkshire, though as the story of Iron Age Yorkshire unfolds, the dependence on archaeological data is supplemented, thinly, by historical evidence; as the Romans encountered the Iron Age peoples of Britain, classical writers were able to record morsels of information. Although iron-working arrived in Britain around 650 BC, it did not instantly launch a technological revolution. Rather, we can detect an intensification of the trends which characterised the latter part of the Bronze Age: the worsening of the climate and the abandonment of water-logged plateaux and valleys; the continuing clearance of woodland and the intensification of farming in the habitable countrysides; and the deepening of the stresses and uncertainties which caused the eruption both of fortifications and of a warrior aristocracy. Whatever form the tribal divisions of the Bronze Age may have taken, we know that, by the time of the Roman invasion, two great tribal confederations had gelled in Yorkshire. The long-favoured eastern parts of our region lay in the territory of the Parisi, a relatively progressive people, while the more rugged terrain in the west formed part of the Pennine realm of the Brigantes, a federation of tough, uncompromising societies with conservative values and a culture which may have owed more to Bronze Age traditions than to Celtic and continental customs; they occupied a territory stretching southwards from the southern uplands to the Trent and Mersey.

Various interpretations of the relationship between the two great tribes of Yorkshire are possible. One of the more recent and most thought-provoking was offered by Higham. He notes that the abandonment of the building of hill-forts in Brigantia took place around 450 BC in the Pennines and a little earlier in the Wolds, though vigorous military engineering continued in the 'zone of insecurity' lying to the north. At the time that hill-fort building came to an end in the territory of the Brigantes, the Arras culture, with its many continental features, had appeared in East Yorkshire, while Yorkshire's western uplands had suffered the worst consequences of climatic deterioration. Higham suggests that the collapse of hill-fort construction marked the establishment of a political hegemony throughout the north of England by the wealthy, influential and powerful bearers of the Arras culture in the Wolds. These people of the east of Yorkshire might have been immigrants from the Continent, but they might equally have imported just the continental cultural trappings. They may have seized slaves and raw materials from Brigantia, exchanging them with members of the 'superior' cultures of Gaul and the Rhineland, and receiving in return the glass, red coral, and metalwork associated with the Arras culture – and importing, too, the continental tribal name of 'Parisi'. Members of this élite tribal grouping may have preserved their distinctive identity by preventing members of subordinate groups from adopting their unique chariot burial rites. Higham suggests: 'The possibility exists that the Arras Culture represents no more than the conspicuous consumption of a community determined to distinguish itself from neighbours and clients and to preserve that distinction by whatever coercion it considered necessary, while funding it in part from the surplus of the economies of these same neighbours.'

During the last four centuries BC, the climate, which had previously severely deteriorated, started to improve; the areas that benefitted most from

Networks of fields worked by Iron Age and Romano-British farmers can still be recognised on the high commons above Grassington. Crops were grown in high places on the limestone, where now only sheep graze and inhabited farmsteads are few

the warmer and drier conditions were the marginal ones, like the Yorkshire Dales. In the period *circa* 200 BC–AD 500, communities in Wharfedale and South Yorkshire responded to the climatic amelioration by establishing enclosed fieldscapes on the better-drained limestones, sands and gravel. This revival of economic fortunes was, according to Higham, translated into military power in the west and north of Brigantia, and the intensive livestock farmers in the territory of the Parisi found themselves unable to resist the challenge from resurgent Brigantia.

If much of this is speculative, the deterioration of climate in the late Bronze Age and early Iron Age is well-established. With the progressive worsening of the climate, population continued to bleed from the rain-lashed uplands and marshy valleys, increasing the pressures on production and possession in the remaining countrysides. Above the 366 metre contour, the Iron Age traveller in the Pennines usually encountered a morass of peatbog. The problems were most severe on the Lancastrian side of the hills, where the ridges intercepted the full force of the depressions wheeling in from the Atlantic. The eastern flanks of the range were fortunate to enjoy something of a rain-shadow effect, and here farmsteads could be found at heights of up to 305 metres. Meanwhile, down in the Vale of York, the slowly rising water-table rendered many low-lying areas no longer habitable, as bog and alder woodland pushed back the frontiers of farming. In such places, grain-fields became damp pastures and pastures surrendered to sphagnum and heather.

As a consequence of these changes, those countrysides that were skill workable had to support the highest possible levels of population, no matter how uninviting their more marginal lands might be. Extensive stands of woodland (apart from the birch and alder woods stretching in ribbons above the waterlogged valleys) became a luxury that society could no longer afford, and they were hacked back, leaving only a minimal amount of economic woodland to serve the demand for building, fencing, tools and fuel. Farming efforts intensified on the more attractive farmlands of the Wolds, the North York Moors, on the drier vales, and on the Magnesian limestone belt between the Pennines and the Vale of York. On the Pennine valley slopes and the eastern flanks of the range, fresh and determined assaults on the wooded zone completed the long phase of clearance in many areas. In Nidderdale, for example, the upper slopes and watershed were completely cleared of woodland by 300 BC, and the birch and alder wood of the lower slopes was diminishing to form a ribbon of woodland along the wet valley floor. In the North York Moors, similarly the Iron Age witnessed the clearance of the last substantial stands of woodland, although strips of woodland still clothed the steeper sides of the valleys which had been cut by glacial meltwater streams. Economically unviable timber could not be allowed to stand here. For thousands of years, the Yorkshire woods had rung to the tune of axes of stone and bronze, but now the iron axes might be thought to have struck with notes both of desperation and finality. Whereas some of the Bronze Age clearances may have been small and sporadic, the clear-felling of extensive areas could have been undertaken in the Iron Age. On Levisham Moor, an ironworker's hut and forge have been excavated; the forge stands on the remains of two earlier forges, so this prolonged period of iron-working would have made heavy demands on the woodland fuel supply.

The system of farming that was practised was a mixed one, including wheat, barley, oats, rye and flax, and animals such as cattle, sheep, goats, horses, pigs and geese. Livestock grazing predominated on the loftier farms and those on the marshy lowlands, while on the better soils the balance swung in favour of cereal crops. Some of the best evidence for cereal growing comes from the discoveries of the stone querns used for grinding the grain. Their distribution pattern shows that arable farming had advanced as far as the upper limits of the area under arable farming at the end of the Second World War, prior to the distorted patterns of cereal cultivation caused by the Common Agricultural Policy. On the higher ground, however, woodland clearance for ploughland or pasture exposed the land to the leaching action of the rain; when farming retreated from the sodden, impoverished soils, it was heather rather than woodland which reclaimed the land. The Iron Age people of Yorkshire used to be regarded as footloose shepherds; in 1954, Sir Mortimer Wheeler wrote of folk living on a diet of 'unmitigated mutton'. Now, however, we know that the settled, mixed-farming life was well-established, and a pollen survey at Roxby, on marginal farming land in the North York Moors, shows that, in the late Iron Age, only about one eighth or less of the countryside was wooded. In the Pennines, too, evidence of cereal cultivation has emerged, and Tinsley concluded that: 'certainly cereals were cultivated in the Pennine province by the Celtic people'. Tinsley also considered that human intervention at several stages in prehistory had played the major role in creating the upland moors:

> Climatic deterioration has long been regarded as a prime cause of woodland decline in the Pennine region, playing an important role in the development of the thick blanket and basin peats. However, as an increasing number of palynological studies are allowing a coherent picture of the vegetational history of the region as a whole to emerge, it appears that we must reassess the relative importance of the roles of climate and man in producing the Pennine moors. The evidence for man as the prime factor in deforestation now appears to be overwhelming, though undoubtedly climatic influences were acting in the same direction.

On the North York Moors, according to Atherden, 'the Iron Age period saw the introduction of the technique of clear-felling woodland in this area'. She thought that Iron Age colonisation and woodland clearance had had a greater impact on the uplands than had been realised and had paved the way for the expansion of heather. One can still discern the field patterns of Iron Age farming in some places. Usually, they are quite invisible to the observer on the ground, but where the underlying geology is favourable, the field-boundary ditches may clearly emerge as 'crop-marks' when the countryside is viewed from the air. Excellent conditions for aerial photography are found on the Bunter sandstone to the east of Doncaster, an area carefully photographed and studied by the aerial archaeologist Derrick Riley. Sometimes the enclosures have an irregular plan, with small blocks of roughly rectangular fields occurring without a clear-cut pattern; sometimes the fields radiate out from the nucleus of a prehistoric farmstead. The most widespread type, however, is the 'brickwork' pattern, where the rectangular fields are arranged like bricks in a wall, with long parallel boundaries and short, staggered cross-links. It is hard to imagine how such networks could have appeared accidentally, so one must conclude that, estate by estate, great tracts of the countryside were divided up in a systematic manner. Aerial photographs may tell one little about the age of the features revealed, but in a couple of places Iron Age pottery of the third or fourth centuries has been recovered from the ditches of brickwork-plan fields. There must be good reason to suppose that other farming areas in the region were similarly divided and that hedgerows originally followed the ditch-lines recorded in the aerial photographs. So one can imagine the working Iron Age countrysides with myriad small, rectangular, hedged fields

stretching from horizon to horizon, with scattered farmsteads standing in hedged, ditched or banked enclosures, and with small stands of economic woodland dotted here and there and ribbons of birch and alder forest tracing the line of the watercourses. A possible exception to this interpretation was the Vale of York, where the clay soils, unlike the Bunter sandstone or the chalk of the wolds, reveal very little to the aerial camera. However, several recent rescue excavations have established the presence of farming and settlement here – for example, when the A19 bypass was built around Easingwold, and the soil-stripping revealed a whole landscape of roundhouses and field ditches with associated mid-to late Iron Age pottery.

In just a few places, the fields where the Iron Age peasants toiled are still visible to the observer on the ground as earthworks or tumbled walls, so that one can walk the old droves and use the mind's eye to re-create countrysides which existed

The boundary banks and tumbled walls of Iron Age and Romano-British fields are plainly visible on the land below Malham Cove

more than 2,000 years ago. The most attractive of such places is found on the upland pastures of High Close and Lea Green, above Grassington in Wharfedale. Here the 'living' stone walls, lead-mining earthworks and traces of medieval farming are superimposed upon the banks and walls of the extensive network of small, rectangular Iron Age and Romano-British fields which pattern about 300 acres of the thinly soiled limestone slopes and plateaux. Rambling across this bare, windswept place, one can easily recognise the hollow ways of the droves, where beasts moved between field and farmstead, and sometimes one can recognise the walled enclosures containing the small, circular dwellings of the Brigantian farmers – one excellent example, Roman in age but Iron Age in style, can be seen on Lea Green above Grassington. Wharfedale has yielded many examples of stone querns, showing that here again there was a mixed-farming economy, with the cultivation of oats and rye to supplement the products of livestock farming. In places where, today, one may meet only a few ramblers and fewer farmers, there was, in the later part of the Iron Age, a substantial population of peasant families. Other patterns of Iron Age fields can be seen in the celebrated view from the heights of Malham Cove, or in Wensleydale, on the flanks of Addlebrough and Penhill. In such places, settlements were established at heights of above 300 metres, and some of the hamlets and farmsteads persisted into the Roman era. Traces of Iron Age dwellings can be encountered in several other parts of the dales, though the evidence is far clearer on the limestone than on the sandstone; the 'hut circles' on the western flanks of Arngill Scar and on Harker Hill in Swaledale are probably of this period, and field patterns similar to those at Grassington are displayed above Calvert Houses in this dale. In the lowlands, the dwellings were of timber rather than of stone and no relics are visible from the ground. Archaeological work shows that the circular farmsteads were often dotted along the roadsides, standing in rectangular enclosures which ran back from the road.

In contrast to the character of the relics from former periods of life in Yorkshire, traces of Iron Age farming are more prominent than the monuments to the contemporary religions. As described by the classical writers who recorded events around the time of the Roman conquest of Britain, Celtic tribesmen worshipped a host of different deities and revered natural features like sacred groves and lakes. Archaeology has shown that the temples of the time were modest, flimsy structures and that there was also a cult for worshipping heads represented by various discoveries of sinister carved stone heads. From at least late Bronze Age times, a water cult seems to have had a prime position amongst the panoply of beliefs; many of our finest museum pieces of the last prehistoric millennium were votive offerings which were cast into bodies of water. In Yorkshire, the best evidence of the water cult is represented by river names and holy wells. The name 'Ure', closely associated with pre-Celtic ritual monuments like the Thornborough henges, could derive from the Celtic *isura*, meaning 'holy one', and the name 'Don' means 'the river of the river goddess'.

Other parts of England can boast many other river names of this kind, but, where holy wells are concerned, Yorkshire has a unique endowment. In 1893, R. C. Hope listed some 67 examples from Yorkshire in his book *Holy Wells of England*, Cornwall came second with 40 wells, and Shropshire third with 36. Even so, Hope greatly underestimated Yorkshire's endowment, listing, for example, only 5 holy wells in Nidderdale, where at least 19 sacred wells actually existed. Although these wells now have Christian dedications, it is plain that the early missionaries pragmatically commandeered the pagan holy places and assimilated them into the Christian tradition. In this way, St Chad's well, at North Ferriby in the lands of the Parisi, has yielded not only medieval offerings, but also votive goods of the late Iron Age. Further west lay the vast territory of the Brigantes, where Brigantia herself, goddess of water, healing and fertility, may have been the principal deity, and where (as E. S. Wood suggested in 1949) most of the holy wells may originally have been dedicated to the goddess.

It is becoming increasingly apparent that Iron Age life in Yorkshire had a maritime as well as a rural dimension. In 1986, excavations began on an Iron Age trading port which stood at Redcliff, on a commanding position where a glacial

The glacial lake of Semer Water, in Ryedale, near Bainbridge. Traces of an Iron Age village and causeway have been recognised here, and there is also a dubious local legend of a more recent village being overwhelmed by the rising waters of the lake

moraine formed the north bank of the River Humber. It appears to have flourished in the first century AD, at a time when Roman forces occupied London and when the frontier of the Fosse Way brought the Roman realm virtually to the riverside here. The Humber was also the boundary between the Parisi and their southern neighbours, the Coritani, and Redcliff's existence as a cultural bridge must have stimulated trade. Crowther wrote: 'From the 40's to the 70's AD traffic in prestigious items like fine imported pottery and brooches may well have been stimulated by the new commercial and diplomatic situation. After the Romans crossed the Humber and pushed northwards in 71 AD, Redcliff was eclipsed by the development of the military and civil centre at Brough (Petuaria) four miles upstream.' Much is known about Roman ships; far less about the craft used by the indigenous people of Iron Age Yorkshire. In 1984, however, archaeologists engaged in the excavation of a glass kiln at Hasholme in the East Riding of Yorkshire noticed some blackened timbers nearby. This resulted in the discovery of a dug out boat, dated to the late or Roman Iron Age, which had been hollowed out from an oak log and lengthened by the addition of extra timbers. It was rather punt-like in form, and had been lost when navigating an inlet off the Humber. It appeared to have been carrying a cargo of unfinished, rough-worked timbers and jointed meat.

Although it has been conventional to assume that the Iron Age witnessed the wholesale conquest of Britain by waves of Celtic invaders from the Continent, the archaeological evidence for such a conquest is modest. It is true, that by the time of the Roman conquest, most, if not all, of England spoke one or other of the Celtic languages, but it is not at all clear that these languages were introduced or imposed by large numbers of Iron Age invaders. In the case of the Brigantes, a thin veneer of Celtic culture may have rested lightly on a people with Bronze Age roots and traditions. On the other hand, the best (if, as has been shown, not uncontroversial) northern evidence for Celtic settlement comes from parts of Yorkshire lying to the east of the Brigantian territory. Around the fourth or fifth century BC, settlers or influences coming from the Marne region of France and associated with a well-developed Celtic culture named after an important archaeological type-site at La Tène on Lake Neuchâtel, Switzerland, became established on the east coast of Yorkshire. In due course, the country of the Wolds and Derwent experienced an eruption of ritual monuments – in the form of square barrows – which is virtually unique to the region.

Hundreds of these square barrows are known to exist, and remarkable information about their

The well-drained limestone countrysides still supported agriculture during the moist climates of the Iron Age. In some places, settlements were found at heights of 518 metres, and one existed just below the 477-metre summit of Addleborough

contents has come from excavations at the adjacent gravel-quarrying sites at Wetwang Slack and Garton Slack, a few miles to the west of Driffield. In 1984 three chariot burials were discovered when quarrying removed the covering barrows. In each case, a chariot had been dismantled and the wooden wheels, with their iron tyres and naves of bronze or iron, were placed side by side on the floor of the grave, and a crouched corpse, lying in a square structure – perhaps the body of the chariot – was placed upon the wheels. The richest burial was that of a woman, who was provided with bronze horse-bits and five 'terrets' (rings fastened to a yoke through which the reins passed); these were of bronze, ornamented with coral imported from the Mediterranean, while the harness fittings in the adjacent warrior graves were less ornate and made of iron. The noblewoman was provided with a side of pork to nourish her in the afterlife, a circular iron hand-mirror, a dress pin and a bronze workbox. The two warrior burials lay to the north and south of this tomb. In one case, it was found that the man was buried with his sword and scabbard, seven broken spears and a shield, while the other warrior had a similar sword and a shield. These were not the first Yorkshire chariot burials to be excavated, but the first modern excavations of this type to produce military grave-goods. Previously, the lack of armaments had led excavators to conclude that the buried vehicles were not war chariots, but light farm carts. One can only wonder about the funeral ceremonies which preceded the burials and the raising of the distinctive square covering mounds, but it is possible to suppose that the corpse was brought from the mortuary house to the grave in a chariot, which was then solemnly dismantled at the graveside. Also, the confinement of the chariot burials to an area to the east of York suggests that the Brigantes – Parisi division of Yorkshire already existed and that the ancestral Brigantes had rejected such novel and alien rites – or else, according to the alternative scenario mentioned above, that they had been prevented from imitating them.

One might again be in the position of knowing more about the dead than the living were it not for other excavations at Wetwang Slack which took place in the 1970s. Although the rite of chariot burial, with its parallels in some continental Celtic communities, provides the best evidence of an Iron Age invasion, it emerged that the Iron Age people of East Yorkshire lived in dwellings that were indistinguishable from those used by other lowland communities in England at that time: circular timber houses with conical roofs of thatch carried on upright posts. These houses had diameters of 9–12 metres, with walls of upright timbers set edge to edge and standing in a circular slot. Their entrance porches faced to the south-east, away from the prevailing wind. Storage pits were dug in the floors of the dwellings, though grain was probably also stored in raised granaries supported by four or six posts. Excavated dwellings that revealed no traces of occupation may have served as mortuary houses. Roadways, some of them defined by ditches, traversed the well-populated countryside, and hamlets and their cemeteries might have occurred every mile or so along the track.

The Iron Age settlement at Wetwang lay inside a long earthwork which enclosed an area of almost 0.7 square miles (2 square km), and its barrow cemetery stretched along the southern edge of the enclosure, which followed the road. At the beginning of this period, roundhouses were scattered along the valley at distances of 45–92 metres, but later the dwellings tended to nucleate in the central area of the enclosure, while the barrows tended to diminish in size, so that burials could only be covered properly by placing the corpses in deep pits. Towards the end of the Iron Age, the barrow cemeteries were abandoned and the clustered settlements expanded. The barrow cemeteries were replaced along the roadside by square enclosures, which could have had ritual uses or simply have been pens for livestock. Excavation of these enclosures has revealed numerous figurines carved from chalk and then broken, raising questions about whether the smashing and decapitation of such figures was a rite which had overtaken the outmoded barrow rituals. The figurines continued to be made after the Roman occupation, well into the second century AD, when settlement drifted away from the site.

New evidence of the rituals associated with the Yorkshire chariot burials emerged in 1986, during

Crop-marks in this photograph of a site in Ryedale reveal the outlines of two (presumably) Iron Age circular houses standing in a rectangular enclosure which seems to overlap with an older rectangular enclosure (the marks in the upper part of the photograph are caused by wind and rain damage to the growing crop).
(North Yorkshire CC Archaeology Section)

excavations at Garton Station in the Wolds. When a barrow was opened, a dismantled chariot was found, its wheels stacked against the side of the burial pit and preserved in the form of clay, which had washed into the grave and replaced the wood as it rotted. Nearby were ditched enclosures, later used for Anglo-Saxon burials, but possibly serving as Iron Age temples associated with the barrow cemetery. Also discovered were round barrows and the remains of a most remarkable funeral ritual. In one example, the body of a warrior was placed in a crouched position, his sword behind his back. Seven spears were then thrown at the corpse and, as the grave was filled, seven more spears were hurled, the heads of which were discovered by the excavators at different levels in the grave-filling. In all four of the round barrows explored, a single body was found together with spearheads, as many as 11 in one case. The excavators suspected that they had unearthed the evidence of a ghost-killing ceremony, performed to ensure that the spirit of the dead would not return to haunt the living. The round barrows were also placed beside the course of the Gypsey

Race, a stream which rarely flows above ground here, and the possibility of a water ritual was also noted. The barrow cemetery was also associated with a double-dyke boundary earthwork, an estate boundary which adjoins other such cemeteries. Subsequently, Stead explored another chariot burial in an adjacent field just across the parish boundary in Kilburn. This was unique because the warrior had been buried in a complete coat of mail, the earliest such armour to be discovered in Britain. His corpse had been placed upon the wheels of his dismantled chariot and the coat of mail was placed, inverted, on top of him. Then the chariot was placed upside down over the body. Most strangely, an adjacent barrow was found to contain the burial of an Iron Age woman and her newborn child; above her, the burial of a pregnant Anglo-Saxon woman had been inserted into the Iron Age barrow. Another round barrow with evidence of the 'ghost-killing' ritual was found, three spears having been hurled into the grave of a warrior who was buried with a spectacular sword which was inlaid with red coral.

The picture that is building up of Iron Age Yorkshire is one of old countrysides which were kept hard at work and fiercely defended. Scores of the winding lanes that are still used today probably date from these times, though the ancestors of most surviving villages had not yet formed. Farmsteads and hamlets heavily stippled the scene; although the archaic tendency for settlements to be short-lived and to drift from one site to another still prevailed, there was also a tendency for larger groupings of farmsteads to appear. Farming of a mixed type was still the norm, although sheep were more widespread than before. Livestock wintered in the open, and the winter sowing of crops had become popular. Although livestock farming may have predominated in many parts of the region, excavations at an Iron Age settlement at Ledston, near Leeds,

have revealed many storage pits of the type used for storing grain. However, there was more to Iron Age life in Yorkshire than wholesome peasant endeavour, and the insecurities and the centralised leadership of the age are evidenced by the construction of hill-forts, large and small.

A hill-fort could be several things: the defensable bolt-hole of a threatened community; a fortified village or 'town'; the place where a chieftain gathered together the livestock and grain of the fiefdom, taking tribute and redistributing the surplus as largesse; or, perhaps, a communal focus for trading, assemblies and ritual. It could also be various of these things at once. In the Brigantian lands, a number of impressive strongholds were built. On the shield-shaped summit plateau of Ingleborough, at a height of 722 metres, the loftiest of all British hill-forts was created. A single stone wall encircled an area measuring about 329 metres by 229 metres, and corresponding to virtually the entire summit area. Within the modest shelter of this wall, about twenty circular stone buildings were erected, although it is hard to imagine that they could have been occupied when winter gales swept the bare summit. Most probably, the dwellings, which were clustered in groups, were occupied during the summer, when flocks or herds grazed the high pastures as part of a transhumance cycle. A recent survey has demonstrated that the method of rampart construction employed here was unique in England.

Ingleborough. The area of the summit, here wreathed in clouds, was the site of Britain's loftiest hill-fort

It was a gritstone wall about 3–5 metres thick, faced on the inner side with large, boulder-like orthostatic blocks and, on the outer side, with courses of drystone walling. Nowhere are the faces well-preserved, and the maximum surviving height of the walling is six courses. The problem of destruction is worsened by ramblers, who take stones from the rampart to add to the summit cairn. Material for the rampart was scooped from shallow quarries just inside and outside the ramparts, with a berm of about 10 metres separating the wall from the summit-edge break of the slope. The internal structure of the wall was based on through-stones, set on edge to divide the rampart into 'boxes' which were filled with rubble, with lines of throughs occurring about every two metres.

Ptolemy, the Roman geographer, listed nine Brigantian 'towns': two of them have not been identified with certainty; five became Roman fort sites – Aldborough, Ilkley, York, Catterick and Binchester – and two – Ingleborough and Stanwick – did not. While Ingleborough has the most imposing site of any Brigantian hill-fort, Stanwick, near Richmond, is by far the most remarkable construction (but since it appears to date from after the Roman invasion, it is reserved for the following chapter). It is worth noting that, with Ingleborough and Stanwick, Yorkshire can boast both the loftiest and the largest of British hill-forts. Ingleborough stands amongst scenes of fine upland limestone scenery, but there are other important Yorkshire hill-forts in more populous settings: Almondbury in Huddersfield has already been mentioned; and Wincobank stands in the industrial heart of Sheffield. Its oval enclosure was guarded by a rampart, ditch and counterscarp bank; the rampart was built of rubble, reinforced by timbers and a drystone revetment. During a conquest or an unexplained ritual, some of the lacing timbers were ignited, turning the adjacent stones to glass. Forts that have been vitrified in this way are common in Scotland, but rather unusual in England. Not far from Sheffield is another large fort, Carlwark, perched high on Hathersage Moor. Stone walls linked the craggy outcrops which hem the promontory, and a great wall, some three metres high, guarded the western approaches. Near Wincobank, the Romans built

Bank Slack, near Harrogate, where the ramparts of the small Iron Age hill-fort rise above the fronting ditch, which now accommodates a stone field wall

their fort at Templeborough, just across the Don, perhaps seeking to police and neutralise the hill-fort – while at Barwick in Elmet the Norman motte and bailey commandeered a great hourglass-shaped fort and exploited its ready-made defences. In addition to these important citadels, there are also many lesser Iron Age strongholds in Yorkshire, like the double-ramparted promontory fort of Bank Slack at Fewston, near Harrogate; the small stone-walled enclosure above Grass Wood, near Grassington; and the forts which stood at Knaresborough and at Laverton, near Ripon.

Hill-forts are monuments to the power structures and insecurities of their day; as in previous times, the need to protect tribal or communal frontiers was also expressed by the construction of long stretches of linear banks and ditches. Yorkshire has many examples of these linear earthworks, although very few have been dated with any certainty. Roman Rig might be a frontier work built by the Brigantes to guard against incursions by their southern neighbours, the Coritani, or even a hopeful shield against the advancing Romans. It runs from Sheffield, following the northern side of the Don as far as the River Dearne. At Wincobank it forks, one branch running to Mexborough, and the other passing through Greasbrough and Upper Haugh. On the western margins of the Vale of York there are several sets of similar earthworks, generally assumed to date from the Dark Ages; the Aberford dykes are unlikely to belong to the Iron Age. A variety of frontier works appeared in the North York Moors like the multiple ditches of the Scamridge dykes, cutting across the county between Trouts Dale and Kirk Dale. Also in the fields above Ebberston are the Oxmoor and Cockmoor dykes. Spratt, who thought that the Yorkshire dykes date from 1000 BC onwards, wrote that the splendid Scamridge dykes:

> six abreast, sweeping a huge arc from the scarp near Cockmoor to the head of Kirkdale, Ebberston, might well have originated not as a local demarcation, but as a major tribal boundary. It was only later converted to a more typical local boundary by the interposition of the Netherby Dale Dyke. When there is a huge earthwork which does not seem to make

sense, in terms of equitable division of farm land, then it probably had a political significance to large numbers of people.

In the Wolds, the country was partitioned by the late Bronze Age and early Iron Age entrench-

An Iron Age site near Malton, revealed by infra-red aerial photography. A pattern of small Romano-British fields is shown, with the fields respecting the ancient track which runs from the bottom centre towards the centre of the picture and then bends sharply to the right. The fields, however, cut across the circles, which represent Iron Age houses, and are also known to cut the Iron Age square barrows, which can just be discerned in the centre of the bottom modern field. At the very top of the photograph, a faint oval feature which is cut by later fields may be a Neolithic long barrow

ments; in the Pennines, a variety of linear earthworks appeared. The most impressive is the lofty Tor Dyke, which severed the routeway which is now represented by the Kettlewell to Coverdale road. It has been suggested that it was built to check the Roman advance on Stanwick hill-fort, but there is no dating evidence to back up this claim and it might simply represent a parochial attempt to keep Wharfedale people out of Coverdale. Just to the north of Richmond and sweeping north in the direction of Stanwick fort is Scotts Dyke, probably an Iron Age boundary embankment.

As the last of prehistoric time trickled away, the people of Yorkshire could look out across a landscape that had been tamed and transformed to meet the needs of the ploughman, stockman and shepherd. Traversed by lanes and trackways and criss-crossed by hedges, the countryside was divided into large estates; on the higher slopes, the stone-walled fields marched to the limits of the farmable area. Smiths and merchants followed the lanes from hamlet to hamlet, perhaps visiting the strongholds and palisaded homesteads of the nobles – places which served as focuses in a land which still lacked recognisable towns. Were one able to transport a modern country dweller from the Dales or moors back in time to the Iron Age, they would probably recognise their localities quite easily. Many of the boundaries and most of the lanes would already be in place, though the absence of contemporary villages or their ancestors would be a striking difference.

REFERENCES

M. A. Atherden, 'The impact of late prehistoric cultures on the vegetation of the North Yorkshire Moors', *Transactions of the Institute of British Geographers*, NS 1 (1976), pp. 284–300.

M. C. B. Bowden, D. A. Mackay and N. K. Blood, 'A new survey of Ingleborough hillfort, North Yorkshire', *Proceedings of the Prehistoric Society*, 55 (1989), pp. 267–71.

D. Crowther, 'Redcliff', *Current Archaeology*, 104 (1987), pp. 284–5.

J. C. Dent, 'The impact of Roman rule on native society in the territory of the Parisi', *Britannia* 14 (1983), pp. 35–44.

J. C. Dent, 'Wetwang: a third chariot', *Current Archaeology* 95 (1985), pp. 360–1.

N. J. Higham, 'Brigantia Revisited', *Northern History* 23 (1987), pp. 1–19.

P. Halkon, 'The Holme Project', *Current Archaeology*, 115 (1989), pp. 258–61.

A. King, *Early Pennine Settlement* (Clapham: 1970).

A. King, 'The Ingleborough hill-fort, North Yorkshire', *Yorkshire Archaeological Society Prehistoric Research Section Bulletin*, 24 (1987).

S. McGrail and M. Millett, 'Recovering the Hasholme logboat' *Current Archaeology*, 99 (1986), pp. 112–13.

D. N. Riley, *Early Landscapes from the Air*, (Sheffield: 1980).

D. A. Spratt, 'The prehistoric remains' in D. A. Spratt and B. J. D. Harrison (eds,) *The North York Moors* (Newton Abbot: 1989), pp. 28–44.

I. M. Stead, 'Yorkshire before the Romans: some recent discoveries' in R. M. Butler (ed.), *Soldier and Civilian in Roman York* (Leicester: 1971), pp. 21–44.

I. M. Stead, *The Arras Culture* (York: 1979).

I. M. Stead, 'Garton Station', *Current Archaeology*, 103 (1987), pp. 234–237.

I. M. Stead, 'Kirkburn', *Current Archaeology*, 111 (1988), pp. 115–117.

H. Tinsley, 'Cultural influences on Pennine vegetation with particular reference to North Yorkshire', *Transactions of the Institute of British Geographers*, NS 1 (1976), pp. 310–22.

York Archaeological Trust, 'Easingwold' *Current Archaeology*, 140 (1994), p. 308.

CHAPTER SIX

The Roman Empire and Yorkshire

THE ROMAN CONQUEST OF YORKSHIRE after AD 69 was largely the consequence of a political scandal which threatened to destabilise the political order in Britain. In the summer of AD 43, a massive Roman invasion force had landed at Richborough, Chichester and at Deal or Reculver, and the southern tribes were soon defeated in a prolonged and bloody battle at a ford on the Medway. The Romans had come to bolster the precarious reputation of their ineffectual emperor, Claudius; to control Britain's important agricultural, human and mineral resources; to suppress the Celtic agitators who harboured dissidents and spread unrest amongst the Celts of the continental empire; and, perhaps, in the eternal imperialist tradition, to bring enlightenment to the pagan barbarians. The invasion revealed all the frailties of British political life: the disunity of the tribes and the jealousy and treachery of their rulers; the indiscipline and antiquated arms and tactics of the tribal armies; and the lack of any sense of island-wide nationhood. It also nipped in the bud the first and only real hope for a home-grown civilisation; given just another century for urban and commercial development, the story might have been different.

By AD 47 a frontier had been established which stretched from Humberside to Lyme Bay, bisecting Britain diagonally along a line which came to be marked by the Fosse Way. Thereafter, the resources of the legions and auxiliaries were engaged in the occupation of the south-west of England, the pacification of the Welsh tribes and, after AD 60, the brutal suppression of the terrifying revolt by Boudicca's Iceni in Norfolk. In the north of England, Queen Cartimandua of the Brigantes had displayed a canny grasp of the *real politik* of the times, putting aside any sympathies that she

might have felt for the 'British' cause and accepting terms which cast Brigantia and the subordinate territory of the Parisi in the role of a semi-autonomous client kingdom of the Roman empire. And so Brigantia waited quietly on the sidelines as the disciplined legions swept through the tribal armies of the southern and western tribes. During AD 50, Roman forces intervened to support the queen against a rival faction and, in AD 51 Cartimandua demonstrated her reliability by handing over the guerrilla leader Caratacus, who had sought asylum amongst the northern tribes. In AD 69, however, she showed that her heart could overrule her head by divorcing her husband Venutius in favour of his armour-bearer. Venutius responded to this grinding humiliation by launching a palace revolution and rallying the Brigantian 'nationalists' to challenge Roman rule. Hitherto, the Romans had been reluctant to contemplate the annexation of the difficult upland territory of Brigantia. Now, however, they were unable to countenance either a threat to their northern flank or the loss of face incurred by the desertion of a loyal client. Vespatian, newly emerged as the victor in a four-way contest for the imperial throne, despatched the IX Legion, with auxiliary support, to the north under the leadership of Petillius Cerialis. Cartimandua was to be rescued and the insurgents suppressed – but this time the old relationship with Brigantia was abandoned in favour of a military occupation. The ultimate fates of the queen and the rebel leader are uncertain.

From Lincoln, the army crossed the Humber and marched through the territory of the Parisi to Malton, swinging westwards along the Vale of Pickering and establishing a legionary fortress at York. Meanwhile, it was believed, the Brigantian

resistance had been focused on the great hill-fort at Stanwick near Richmond. In 1951–2 the stronghold was excavated by Sir Mortimer Wheeler, at a time when it was still fashionable to formulate theses to link such places to recorded battles and presumed invasions. Not surprisingly, the maestro identified the central defensive unit, a fortified hill known as 'the Tofts', as the citadel of Venutius and his supporters, established after the fracas of AD 50. About a decade later, it was argued, a great outer rampart was added to the north of the Tofts, enclosing some 130 acres with a ditch that was 4.8 metres deep by 12.2 metres wide, and a rampart of earth and rubble faced in drystone walling. Finally, around the time of the Roman penetration of Yorkshire, a further 600 acres to the south of the Tofts were enclosed by another great loop of bank and ditch. According to this interpretation, while Cartimandua and the Roman faction presided in the south of Brigantia, perhaps ensconced in Almondbury hill-fort, Venutius was preparing his capital and fortress at Stanwick. More recent interpretations sound a somewhat different and rather cautious note. Instead of being a purpose-built nucleus for rebellion, the hill-fort is seen as an established capital and trading centre which engaged in a lucrative marketing relationship with the occupied territories to the south and which was evacuated rather than re-fortified as the invasion forces drew near.

According to Higham, Wheeler's interpretation is unacceptable. The fort was a gigantic construction, the site being enclosed by banks that were partly faced in stone and by ditches hewn into rock in a way that required the removal of some

The ramparts of the enormous British fortress at Stanwick are still prominent features of the countryside

300,000 cubic metres of earth and rock. As much a town as a fort, Stanwick was an *oppidum*, the only such capital in northern Britain and an enormous project which advertised the coercive power of its proprietors; it existed to make a political statement to other British people. Its location was a crucial one in terms of north–south relations and the control of the North of England. With all this in mind, Stanwick is attributed not to Venutius, but to the Roman puppet queen, Cartimandua herself. It seems to have existed at the same time as her kingdom and ended around AD 70; as the nerve centre of the North and close to the territories occupied by subordinate tribes, it was probably the normal location for her household. Stanwick's poor access to navigable waterways made it an indifferent centre for commerce, but it could be regarded as a centre for consumption and a place where Brigantian slaves and raw materials could be concentrated for export to the south. Higham even raises the possibility that Roman troops were stationed at Stanwick and may have been involved in the rescue of Cartimandua in AD 69.

During the second half of the 1980s, new excavations were undertaken at Stanwick which supported the identification of the *oppidum* as Cartimandua's capital. The order of construction proposed by Wheeler was reversed, so that, instead of beginning small and growing larger, it

An excavated and restored section of the Stanwick fortifications, showing the imposing nature of the defences

began large. The outer circuit of defences was built first, perhaps developing from existing estate boundaries, and then about a fifth of the interior, containing the main settlement focus, was defended by massive earthworks. Settlement at the Tofts went back to the middle Iron Age, but the defences here came late in the sequence and did not constitute a primary defensive enclosure. Rather, they resulted from a diversion of a main defensive earthwork to provide protection for the existing settlement. The excavations revealed that the occupants were not simple shepherds, but mixed farmers who grew spelt wheat and six-row barley and who imported Roman products on a grand scale. These included drinking vessels, wine and oil, and Haselgrove writes that:

Many of these imports are extremely unusual: rare Samian forms, uncommon amphora varieties, and volcanic glass hardly known outside Italy; they have the look more of diplomatic gifts than normal trade goods. They mostly date to the short period between the Claudian invasion of south-east England and the Roman conquest of northern England soon after AD 70. The imports then, like the earthworks, mark out Stanwick as a settlement of unusual distinction and importance . . . it could have been the seat of an important pro-Roman ruler, for the imposing but essentially impossible-to-defend fortifications show no traces of having ever been attacked or slighted by the Roman army. Might we then do better to view them rather as a statement of the exceptional standing of whoever commissioned them, their purpose above all to impress dependents and visitors alike?

Stanwick seems to have seen major redevelopments in the middle of the first century AD, with enormous earthworks being built, perhaps to separate private and public spaces. But its prestige seems to have been tied to the person and patronage of Cartimandua; when she fled, it was rapidly abandoned.

In the event the conquest of Yorkshire did not overtax the Roman military machine, but, having been drawn into the occupation of Brigantia, new questions about the north-western limits of the empire were raised. A series of marching camps punctuate the line of advance westwards to Stainmore Pass and beyond, to the Vale of Eden and Carlisle. Between AD 78 and 80, the new governor, Julius Agricola, engaged in an invasion of Scotland which reached northwards to the Tay; the situation was changeable in the years that followed, and various defensive lines were explored in the far north. Then, following a visit to Britain by Emperor Hadrian in AD 122, plans for the building of a great wall were launched. With the completion of Hadrian's Wall in AD 130, Yorkshire lay fairly securely within the Roman fold. Brigantian unrest in the Pennines remained, for most of the time, no more than a minor irritation, though there were rich lands and the basic prerequisites of civilised life to the east; here the two forts at Malton and Brough-on-Humber were deemed sufficient to subdue the lands of the Parisi.

The first phases in the Roman occupation of Yorkshire were military in character, and a series of fortified auxiliary camps were built as the infantry and cavalry units probed the Pennine territories of the Brigantes. During Agricola's governorship, permanent fortresses were constructed to accommodate the army of occupation at sites that includes Bainbridge near Hawes, Elslack near Skipton, Ilkley, Bowes, Newton Kyme near Tadcaster, Castleford and Catterick. Such forts, designed to accommodate garrisons of 500 or 1000 troops, were typically constructed to a playing-card-shaped plan, their rounded corners being more suitable for defence. An outer ditch and a rampart of earth protected the commandant's house, the headquarters and the granary – buildings, which were placed at the heart of the camp – while neat rows of long barracks lay to either side of them. Some of the forts experienced a long occupation, during which the lay-out of buildings might change and the ramparts might acquire stone walls. At Castleford, both the early Roman fort, *Lagentium*, and a late Roman fort

have recently been excavated in the area of All Saints' Parish Church, with great ditches being cut across the earlier defences when the site was refortified in the third or fourth century. During the course of their long occupation, the Romans built some forty forts in Yorkshire, mainly in the west, though few survive as prominent monuments today. The fort at Bainbridge is now seen as a low rectangular platform crowning the rounded hillock, a drumlin, which overlooks the village. To the Romans, it was *Virosidium* or 'High Seat'. It began as an Agricolan fort with earthen ramparts and timber buildings which was established deep in Brigantian territory. Roman troops were stationed here until the end of the fourth century; stone fortifications were built and rebuilt on different occasions, and eventually a native village and, perhaps, a market grew outside the walls. The first refurbishment of the defences could have been carried out in the aftermath of a great northern rebellion, which flared southwards as far as Derbyshire in AD 155, although inscriptions referring to governors appointed under Emperor Severus also indicate repair work following an invasion in AD 196–7. The frontiers in the mid-second century were probably the result of the premature removal of Pennine garrisons to man the campaign in Scotland, while the later rebellion in tribal lands both north and south of the wall was stimulated by a withdrawal of troops to fight in Gaul. At no time, however, can one imagine *Virosidium* being regarded as an idyllic posting, either by the sophisticated Roman commanders or by the garrisons of men who were recruited from many corners of the empire – all of them warmer, sunnier and much less windswept than Bainbridge. In the Pennines and the North York Moors, hunting provided a diversion for bored troops. Soldiers could not be allowed to grow lax and complacent, however, and at Cawthorn, on the moors near Pickering, they were drilled in fort construction, siege warfare, ballista-firing, the construction of field ovens and the digging of latrines. The earthworks of a series of practice camps and a practice fort can still be seen.

The Romans built great stone-walled legionary fortresses in Britain; places like Caerleon in Gwent, Lincoln, Colchester, Chester and York – the last named being one of the greatest fortresses in the

The Roman fortress of Virosidium *stood on the prominent glacial mound or 'drumlin' which overlooks the village of Bainbridge. Although a native village developed outside the wall of the fort, the modern village may only be able to trace its lineage back to a medieval forest settlement said to have been created in 1227 to accommodate 12 foresters*

entire empire. There is a remote possibility that Roman troops were stationed in York during the days of Cartimandua's client kingdom, and Hartley writes that: 'The general position of the Brigantes as Roman clients under Cartimandua would not necessarily exclude the possibility, since Roman troops were occasionally stationed in client kingdoms in special circumstances. The struggle between Cartimandua and her consort, lasting from the govenorship of Didius Gallus (AD 52–57/8) to AD 69 offers a possible context. But there is no positive evidence for occupation under Gallus . . .'. Like many other Roman towns in England, York began as a military base, its most likely origins being as a turf and timber fortress created by Petillius Cerialis during the initial invasion in AD 71. During the reign of Trajan (98–117) the fort was redeveloped, gaining stone walls, corner towers and gates, the walls superseding

the timber palisades which derived from an earlier refurbishment around AD 81. Meanwhile, a civil settlement was beginning to develop in the protective shadow of the fort. At the time of Hadrian's visit, the VI Legion took over the garrison from the IX Legion, and this legion then remained in occupation throughout the Roman era. The Legio IX Hispana is last mentioned in an inscription of AD 108 and its subsequent 'disappearance' is the subject of much speculation. Some have suggested that it was annihilated in a war with the British, or else humiliated and then cashiered after such an encounter, though Birley suggests that it was transferred to Germany.

Being well along the way to the Scottish battlefields, yet insulated from the fighting by Hadrian's Wall and, for much of the second century, the Antonine Wall, York existed as an excellent base and command post for the Scottish campaigns. In AD 208, the emperor, Severus, used York as his command post; three years later he died and was cremated there. In recognition of York's importance and prestigious role during the campaign, the designated title of *colonia* or 'colony' was granted to the civilian settlement there, the name having previously been used to denote the new towns in which army veterans were settled at the end of their service. Only three other towns in Roman Britain could boast such a lofty status: the *coloniae* at Colchester, Lincoln and Gloucester. York was well established as a provincial capital by this time, a major military and civilian settlement and, as a communication centre, at the hub of the region's road network. Early in the third century the city experienced a comprehensive reorganisation, perhaps coinciding with the promotion to colony status which took place at some point between AD 211 and 237. Roman York had two components, with the Ouse flowing between them. On the north-eastern side of the river, there was the massive rectangular fortress, one of the strongest in the whole empire, its centrally placed headquarters standing on a site now occupied by the Minster. Just across the Ouse to the south-west lay the fortified and neatly planned civilian colony, containing the imperial palace occupied by Severus and, probably, by Emperor Constantius I, who also died at York, and his son Constantine, who was proclaimed emperor there. It is thought that relics of the palace were unearthed when the old railway station was being built in 1839–40. According to Palliser, Roman York should be regarded as a double settlement, comprising the fortress on the left bank of the Ouse and the civilian town on the facing shore. This is generally agreed, but he also believes that this double character persisted into the medieval period and that the Minster had a predecessor on the west bank of the city. He writes that: 'At York, where there was both a *colonia* and a legionary fortress, the suggestion here made is that the Roman and Anglian episcopal church stood in or near the forum of the *colonia*, but that at some unknown date it was superseded by the royal chapel in the heart of the fortress. This probably signalled a shift in importance from west bank to east bank . . .'.

Amongst its other roles, York flourished as a commercial and manufacturing centre, trading with Bordeaux and the Rhineland and supporting important industries in linen and Whitby jet, as well as the less specialised iron, bronze and pottery industries. As yet, the city was undisturbed by a flooding problem and sat astride a bustling waterway used by both local and international shipping; just to the east lay some of the most productive farmlands in the north. Although Yorkshire as such did not exist at this time (though York dominated a territory made up of the old tribal lands that it had gained from the Parisi), the region had gained a prestigious capital, *Eburacum*, the name perhaps deriving from 'Eburos', signifying a man who owned or lived among yew trees (though this was meaningless to English settlers, who interpreted the name as *Eoforwic,*: 'the town of the boar'). The colony of York was a vibrant and cosmopolitan place, home to people of many different classes and a host of different origins. Burial within the Roman town was forbidden; richer residents were buried close to the town, the poorer classes further away. The excavation of one of the poorer cemeteries, at Trentholme Drive, half a mile (0.8km) to the south-west of the colony, has cast some light on life and death amongst the lower-class communities of York. Half the people died between the ages of 20 to 40 and were buried without coffins in unmarked graves; gravediggers would frequently invade recent graves and throw the gruesome relics aside with scant respect for the dead. The people in the working communities were not as short as is often supposed; the women averaged about 1.5 metres in height, while the men averaged a respectable 1.7 metres. Most of the women probably came from the native British stock, but the male skeletons showed considerable ethnic diversity, allowing us to imagine native wives being taken by retired soldiers, traders, settlers and slaves from most of the far-flung provinces of the empire. Most of the skeletons displayed signs of tough physical work.

Around AD 300 it was thought necessary to strengthen the defences of York even further, and a series of imposing and prestigious bastions were added to the riverside wall of the fortress – the Multangular Tower is a celebrated survival from this phase of fortification. England continued to be a prosperous and valuable outpost of the empire, but it was threatened on all sides by mounting threats of barbarian invasion. Were a reminder of the dangers needed, it arrived in AD 367, when the Picts, Scots and Saxons gathered themselves together in a massive co-ordinated raid, taking the barbarians into places which had hitherto been thought secure. York survived the invasion, but began to suffer from another kind of seaborne threat, represented by a rise in sea level

The Multangular Tower at York, seen from the inside. The massive stonework of the medieval wall fortifications stands above the Roman walling. (A collection of stone coffins has been placed at the foot of the wall.)

and a consequent disruption of the river system. Parts of the city lying below the 10.7 metre contour experienced inundations, and sections of the riverside defences, wharfs and harbour installations were destroyed, as well as the vital bridge between the two parts of the town – tragic blows to a centre which relied so heavily on river trade and communications. Excavations have explored the old riverfront, and Ramm writes that:

> it is clear that although the area of the fortress and much of the *colonia* was above flood level, the effect of annual winter flooding on the scale implied by the silting would have been little short of disastrous on a town already weakened economically by the break-up of Roman government and the devastations of raiders . . . The destruction of Roman harbour and wharfage facilities beyond recovery is

illustrated by what happened at Hunngate. In whatever form life continued at York, it was not until the eighth century that she recovered her position as an international trading centre and port.

However, one important legacy of the Roman town would be handed down – the street pattern, which, in its essentials, can still be traced today. Like most other great towns of Roman Britain, York sank into decay, despite the efforts that must have been made to preserve civilisation. Even when the Roman garrison was withdrawn, the dignitaries seem to have employed mercenaries or irregular pagan troops to defend their city, but, in the fifth century, civilisation and settled urban life apparently collapsed.

The other towns of Roman Yorkshire were of a less prestigious and grandiose nature; Aldborough, Brough-on-Humber, Catterick and probably a town near Malton were distinctly provincial in comparison to the splendours of York. Aldborough, near Boroughbridge, was *Isurium Brigantum*, the civilised capital of the Brigantes, the first component of its name probably referred to Isura, the River Ure. Like York, Aldborough

probably had its origin in an invasion fort, which gradually acquired the walls, gates and bastions needed to establish its status and protect its 55-acre interior. A section of the town's red sandstone wall survives, now more impressive for its width – nearly 2.7 metres – than for its height. The town sank into decay at the end of the Roman period, but hints of its civilised urban life are evident in two surviving mosaics, although the best example – showing Romulus, Remus and the she-wolf – was moved to Leeds City Museum. Apparently Aldborough supported its own little 'guild' of mosaic workers.

Brough-on-Humber (*Petuaria*) grew from an abandoned Roman fort lying close to a large native settlement and presumed prehistoric ferry route – perhaps the place where Petillius Cerialis and his legions made their crossing from Lincolnshire. When the fort at Brough was abandoned, in the late 70s a storage depot was maintained and, some years later, the site was developed as the administrative capital of the Parisi. The tribesmen seem to have had little enthusiasm for their capital, however, and, by the third century, it was in decline; the territory was perhaps taken over by York, and it may have failed completely before the

The complicated evolution of York's walls, exposed in an excavated section around the Anglian Tower, looking towards the distant Multangular Tower. The Anglian Tower (see p.89) plugs a gap in the Roman walls; beyond it, the remains of a Roman interval tower project from the right. The wall to the left of the photograph is the twelfth-century boundary wall of St Leonard's Hospital. Above the masonry of the Roman wall, to the right, is the thirteenth-century city wall, now standing on a strengthened concrete base with a cobbled face

more general collapse of urban life. In its heyday, however, the town could boast its own theatre – an unusually prestigious facility for such a small town, and the gift of a private benefactor. *Petuaria* had docks on the Humber and could have served the group of villas lying to the north. Much about the town remains mysterious; the fortified area might have had a purely military and naval function and the civilian settlement could have developed outside its walls, perhaps experiencing an ephemeral existence at or by the nearby settlement at North Ferriby.

Catterick (*Cataractonium*) lacked some of the

A beautiful mosaic floor displayed at the lost Roman town of Isurium Brigantum
(Aldborough, later a rotten borough)

prescribed functions of a Roman town, yet proved to be of more substance than Brough. It, too, developed as a fort site and grew to acquire walls and a planned grid-work of streets which extended across 25 acres and is now cut across by the A1 just to the south-east of Richmond. In 1995 an archaeological discovery of international importance was made when the town's amphitheatre was discovered beneath Catterick racecourse. The Roman town lay to the north of the racecourse, and the Roman fort to the east. The interior of the amphitheatre was about 140 metres in diameter, was surrounded by a broad bank and had a finely cobbled surface. Roman pottery taken from the interior suggested that the amphitheatre was built towards the end of the second century AD. The excavation work showed that the builders had incorporated a large, circular burial cairn of late Neolithic/early Bronze Age date. This tomb was 35 metres in diameter and contained eight small burial chambers; its unusually large size suggests that the Catterick area may have been an important focus of Neolithic life. After the

abandonment of the amphitheatre, an Anglo-Saxon cemetery was established against its south-eastern margin in the period AD 450–550. It is conceivable that some of the bodies might have been those of English warriors killed in the defeat of the Britons by the Anglo-Saxons at 'Catraeth', which is assumed to be Catterick, around AD 600. The excavation of buildings in an adjacent Iron Age enclosure showed that towards or after the end of the Roman occupation, a blacksmith had set up his furnace in the remains of a long-deserted dwelling. Urban life of a kind survived at Catterick until well after the Roman withdrawal – either because of or in spite of the probable recruitment of Germanic mercenaries to protect the threatened settlement. The town appears to have been evacuated, but then resettled by strangers. Wacher writes that: 'This reoccupation, probably now of a civilian character, came before the main waves of Anglo Saxon settlers reached this part of Yorkshire . . . It probably dates to the very end of the fourth or more likely to the early part of the fifth century.' The rebuilding involved timber-framed buildings with sleeper beams set on boulder footings; they were set out in a manner that ignored pre-existing building lines, suggesting that the former lay-out had become overgrown and obscured by then. In addition to the towns mentioned, a number of Romanised townlets or large villages seem to have been established at places that included Leeds, Wetherby and Adel.

A different Roman contribution to the civilian settlements of Yorkshire was represented by the establishment of a number of small country mansions and estate centres, the villas. These were both working farmsteads and country houses; many of them became quite elegant homes, with expensive mosaic floors and central heating systems, during the later phases of Roman rule. Their owners were frequently either native nobles or entrepreneurs who had adopted Roman values and lifestyles or foreigners – retired army officers, political exiles, merchants and the like – or even absentee landlords who lived on the Continent. The villas were scattered amongst the native farmsteads, but they were not evenly distributed throughout the north. Being prosperous commercial farmers, the owners of the villas sought locations with good soils, easy access to urban or military markets or ports, a reliable labour force, and security from local insurgencies and trouble makers. Consequently, they avoided the rugged terrain and less reliable communities of the Pennines and dales, the villas discovered at Gargrave and Piercebridge being unusual stragglers from the flock. Instead, builders of the villas gravitated to the traditionally favoured countryside of the Vale of York and the wolds and tended to develop estates close to one of the towns or to the fort of Malton, which probably had a town nearby.

Some villas commandeered sites previously occupied by native farmsteads, though changes in ownership are not necessarily implied. The early villas could be quite simple, long and narrow rectangular buildings, but as the standards of prosperity rose under Roman rule, they tended to sprout cross-wings or grow around a courtyard. They could be built of stone or of timber-framing on stone footings; roofs could be of thatch or shingles, although clay tiles were often favoured. Nearby stood the barns, stables, granaries, yards and paddocks, and the cottages of the farm-workers. Most of the villa estates in Yorkshire were associated with a mixed type of farming; as well as discovering the bones of animals like

cattle, horses, sheep and pigs, excavations tend to uncover the querns used for grinding corn. Although town life in the fourth century was generally in decline, this was the heyday of the country house, and the profits of farming were lavished on splendid mosaic floors for the living-rooms, new bathhouses, hypercausts to ward off the chill of the Yorkshire winter, sculpture, verandas – and, doubtless, elegant gardens too. The most northerly mosaic discovered so far was found in 1966, during the partial excavation of a villa site at Beadlam, near Helmsley. Yorkshire villa owners may have had far fewer aristocratic neighbours than those who lived in the gentler, villa-studded countryside of the South, but at least York offered sophisticated urban amenities, and the numerous northern garrisons provided a large and reliable market for their produce.

The villas, however, played only a relatively minor role in the field of agricultural production; although evidence of the reorganisation of native field systems may be coming to light in East Anglia, the imperialists were generally quite content to rely on the competence of the traditional farming systems and the efforts of native rural communities. Local unrest occasionally flared into rebellion in the Pennines, but farming was the main occupation even there, and most Yorkshire people accepted Roman rule and actively exploited its advantages. Not only did the occupation forces largely remove the curses of war and insecurity, they also constituted an insatiable market for farming produce: they established towns which were hungry for foodstuffs, fuel and industrial raw materials; they built military roads which were soon available as trade routes; and forged the ports and sea routes which gave access to the vast imperial market for British production. At forts like Malton, Slack (near Huddersfield) and many other places, the garrisons acted like magnets to the aspiring tradesmen, artisans and lower grades of hangers-on who could profit from their contacts with the military. Vaguely Romanised villages or *vici* developed along the roads which carried the quickening trade and, in some places, local industries like quarrying, textiles, pottery and mining took root. Crambeck, about 18.6 miles (30km) to the north-east of York, had a complex of kilns that supplied much of the pottery used in northern England during the later years of the occupation, but decayed after the Roman retreat; Tadcaster and Cleckheaton had metalworking and quarrying industries. An iron-smelting furnace has been discovered at Cantley, in the south of the region. Although the lead reserves of the Pennines may have been worked before the conquest, new workings were opened up to exploit the valuable commodity above Pateley Bridge in Nidderdale, in Wharfedale and in Swaledale – and, in true imperialist fashion, the state claimed ownership of all mineral rights.

Although it may be imagined that, after AD 43, Britain suddenly became populated by Romans, true Romans remained a minority even among the highly cosmopolitan ranks of the occupying forces. In terms of the population as a whole, foreigners amounted to only a tiny fraction: England was

still an amalgam of Celtic and older cultures organised by a thin veneer of Roman officials and military personnel. All the people of Britain, perhaps as many as 6 million souls, had to be fed by the domestic farmers, and, at the same time, the continental empire demanded a share of the produce. On the high slopes of the Dales, native peasants worked in fields like those above Grassington, while the traditional arable fields were farmed to breaking-point – and beyond. Farmsteads grew into hamlets, hamlets became villages, and new villages grew beside the roads and at the sites of industry.

This productivity and prosperity was taking place against a background of rising insecurity; while the legions and auxiliaries might be able to police the countryside, they were unable to meet all the threats from the restless barbarians who nibbled at the margins of the empire, ever probing to obtain the spoils of raiding in the soft, rich interior. Germanic seafarers could emerge from the North Sea mists at any time. Scots from Ireland threatened the western seaboard; and Picts probed the defences of the northern frontier. Yorkshire was not entirely secure against the

The remains of the Roman watch- and beacon tower-cum-block house at Scarborough. The garrison here was slaughtered, presumably by North Sea raiders of the late fourth century.
(North Yorkshire CC Archaeology Section)

Pictish threat, and any force which overran the garrisons on Hadrian's Wall might arrive at York just two days later. The east was particularly vulnerable, some of the richest parts of Yorkshire lying within a day's march of the North Sea shores.

To counter the nagging threat of Pictish or Saxon pirates and raiders, the last decades of the fourth century saw the building of a series of lookout posts between the Tees and Flamborough Head, at Filey, Goldsborough, Huntcliff, Ravenscar and Scarborough. These were manned from AD 369 until the Roman withdrawal, or a little later. The Roman signal station at Filey, the southernmost of the five stations perched on high headlands, was placed on Filey Brigg, one of the most spectacular locations in Yorkshire. It consisted of a tall tower which stood in the centre of a walled enclosure, which was in turn surrounded by a ditch. In 1993–4 a re-excavation of the signal station, rapidly eroding from the receding cliffs of the peninsula, was undertaken. The signal stations were block houses as well as watch-towers, and the garrisons at Goldsborough and Scarborough are known to have been overrun and their inhabitants slaughtered when their towers perished. The square stone and timber towers stood about 30.5 metres tall and were protected by the ditch, battlemented walls, gate-house and angle towers which fortified their small surrounding compounds. A beacon burned at the top of each tower, and warnings of invading fleets could be passed to Roman warships and cavalry forces. At Goldsborough, about four miles (6.4km) to the north of Whitby, the stone base of the compound walls survives, and at Scarborough similar foundations run up to the cliff edge. A more interesting survival is Wade's Causeway on Wheeldale Moor, where the footings of a small Roman road are exposed. The origin and destination of the road are unproven, though it probably formed a link between the garrison at Malton and the coastal signalling stations.

The Romans had arrived in Yorkshire as an invincible civilised power, but the final stages of the occupation were marked by insecurity and decay. Barbarian armies and raiding parties were now able to penetrate the imperial frontiers; although garrisons at Malton, Newton Kyme and

The exposed foundations of a Roman road on Wheeldale Moor, above Pickering. The road is presumed to have served the early-warning system of stations along the east coast

the old Agricolan fort at Burwen's Castle, Elslack, near Thorton-in-Craven, were strengthened to resist the incursions, the end of Roman rule was in sight. The legions and heterogenous auxiliary and mercenary forces were not overwhelmingly defeated in Britain, but in about AD 410 the troops were withdrawn to buttress the crumbling continental empire. What, then, was the legacy of Roman rule in Yorkshire? On the negative side, the legions left behind a population which was swollen by the centuries of peace and economic opportunity, yet completely unprepared for home rule. They also left many countrysides that were crippled by overwork and unrelenting production; soil washed down from exposed winter plough-lands in the Wolds choked the river networks, while, on the hills above, the bare white geological bones protruded through the eroding earth. On the more positive side, the empire had demon-strated the potential of unity and integrated development, sowing seeds of civilisation which

would never be forgotten. It had created a magnificent regional capital which, one day, would rise again, and it had seen the establishment of Christianity, a religion which may have smouldered unextinguished in some corners of Yorkshire until the new era of conversion. Perhaps the most tangible legacy was the construction of a network of Yorkshire roads. Most began as arteries built and used by the military, soon became important commercial routeways, and were then to endure

through all the chaos and decay of the Dark Ages. When King Harold came north to defeat the Norsemen at Stamford Bridge in 1066, he marched up the Roman Ermine Street which linked London, Lincoln and York. Had he wished

The Beggarman's Road, in the Pennines, south of Hawes, which has briefly adopted a section of Roman road, forks away to the left, while the Roman Cam High Road continues towards Bainbridge in the guise of a rough track

to venture further north, he could have followed Dere Street to the Firth of Forth. The Roman roads were well used in medieval times and some are still used today. Many journeys in Yorkshire – like the one from York to Boroughbridge – consist of a series of straight Roman sections linked by curving stretches of road formed in the Dark Ages or medieval periods, when travellers abandoned a section of worn-out Roman highway. Similarly, a driver on the A59 Harrogate–Skipton road veers right and left at Kettlesing Head and picks up a two-mile (3.2km) stretch of the road built by Agricola's troops in their advance from the cavalry fort at Ribchester, in Lancashire. The old road disappears beneath Fewston Reservoir, but it can be traced on Blubberhouses Moor, still running arrow-straight beyond the reservoir. For many years to come, there would be people in Yorkshire who longed for the legions to restore civilisation, profits and the lost comforts and security, yet who lacked the ability to re-create the former ways. But there were others, too, who valued the independence or the opportunity to prosper which the return to lawlessness provided.

REFERENCES

P. Abramson, 'The search for Roman Castleford', *Current Archaeology*, 109 (1988), pp. 43–48.

E. B. Birley, 'The fate of the Ninth Legion' in R. M. Butler (ed.), *Soldier and Civilian in Roman York* (Leicester: 1971) pp. 71–80.

T. Garlick, *Roman Sites in Yorkshire* (Clapham: 1978).

R. Hanley, *Villages in Roman Britain* (Princes Risborough: 1987).

B. R. Hartley, 'Roman York and the northern military command to the third century A. D.', in R. M. Butler (ed.), *Soldier and Civilian in Roman York* (Leicester: 1971), pp. 55–70.

C. Haselgrove, 'Stanwick' *Current Archaeology*, 119 (1990), pp. 380–4.

N. J. Higham, 'Brigantia revisited', *Northern History*, 23 (1987), pp. 1–19.

R. F. J. Jones, 'The cemeteries of Roman York' in P. V. Addyman and V. E. Black (eds) *Archaeological Papers from York Presented to M. W. Barley* (York: 1984), pp. 34–42.

D. Miles (ed.), *The Romano-British Countryside*, British Archaeological Report 103 (1982).

C. Moloney, 'Catterick Race Course' *Current Archaeology*, 148 (1996), pp. 128–32.

D. M. Palliser, 'York's West Bank: medieval suburb or urban nucleus?', in P. V. Addyman and V. E. Black (eds), *Archaeological Papers from York Presented to M. W. Barley* (York: 1984) pp. 101–8.

M. S. Parker, 'Some notes on the pre-Norman history of Doncaster', *Yorkshire Archaeological Journal*, 59 (1987), pp.29–43.

H. G. Ramm, 'The end of Roman York', in R. M. Butler (ed.), *Soldier and Civilian in Roman York* (Leicester: 1971), pp. 179–200.

J. S. Wacher, 'Yorkshire towns in the fourth century', in R. M. Butler (ed.), *Soldier and Civilian in Roman York* (Leicester: 1971), pp. 165–77.

J. S. Wacher, *The Towns of Roman Britain* (London: 1976).

L. P. Wenham, *Excavations at Trentholme Drive, York* (London: 1968).

Yorkshire Archaeological Trust, 'Filey', *Current Archaeology*, 140 (1994), p. 312.

Yorkshire Archaeological Trust, 'The waterfront of York: North Street' *Current Archaeology*, 140 (1994), p. 309.

CHAPTER SEVEN

Darkness and Enlightenment

AT THE START OF THE FIFTH CENTURY, England (with Wales) was an embattled outpost of an embattled empire. In 409, with Roman troops withdrawn to defend the imperial heartland, the British leaders recognised the realities of their situation and declared their independence from Rome. The next year, a letter from Emperor Honorius advised the British to look out for themselves as best they could. At first, the Romanised British landholders and office-bearers would have struggled to preserve the institutions, privileges and values of the empire, hiring uncouth foreign mercenaries to defend towns and estates whose populations had lost the old British enthusiasm for battle. The British were completely unprepared for independence, severed from the empire which had provided their territory's *raison d'étre* and from the trading networks which sustained its economic life. The garrison gone, the administrative system breaking down and its people bewildered and divided, Britain subsided slowly into chaos. The new conditions favoured not so much those who had become rich, specialised and inflexible, but those who were less fortunate, who were hardened by adversity, expected little and could adapt their lives to snatch at the opportunities presented by a land in turmoil.

Even if one takes account of all the political traumas, the economic collapse and the undermining of the communal confidence, it is still difficult to explain the magnitude of the disaster which afflicted England during the centuries after AD 409. Much of Iron Age Yorkshire had been an area of prosperous and well-developed farming, with the mosaic of fields often stretching as far as the eye could see. After the Roman withdrawal, farms, estates and whole countrysides surrendered to thorn and thicket; peasants were hacking fields from the woods which had spread across the ancient farmlands for almost a millennium. Evidence is emerging which suggests that, after several centuries of commercial efforts to supply the insatiable imperial market, many Yorkshire grain lands were exhausted, with rivers choked by the run-off of silt. Even if this environmental disaster is added to the list of ills, however, one is still unable to account for the catastrophes which seem as if they might have reduced the region's population down to the levels of the earlier half of the Bronze Age. It is hard to believe that Britain was not devastated by a sequence of terrible plagues, and epidemics are certainly recorded on the Continent at this time.

The scale of the abandonment and decay would have varied from place to place, affecting the traditionally favoured areas least of all. By late Saxon times, the Wolds seem to have been fully exploited as an area of mixed farming and grazing with few trees; the decay which followed the Roman collapse seems to have been minimal here. However, when one looks at the archaeological legacy of the first centuries of the Dark Ages, it is not the wealth and richness of the remains which impress, but the outright paucity of the evidence for settlement. If people are accustomed or reduced to living in flimsy dwellings of timber, mud and thatch in countrysides which are subsequently subject to ploughing, the main source of settlement evidence is likely to come from the pottery scatters which endure at the sites of the vanished homesteads. During the Roman period, prodigious quantities of well-fired pottery were imported to Britain, some of it magnificently decorated, while excellent home-produced wares were manufactured at potteries like the one at Crambeck. Very little pottery has been found which belongs to the centuries that followed. The black, grass-tempered ware of the pagan Saxons is scarce, likely to be poorly preserved and difficult for searchers to spot in the plough-soil. One is left to wonder whether the British, after centuries of familiarity with fine Roman products, actually lost the ability to fire good pots, or else reverted to the use of coarser wares, such as their distant Iron Age ancestors had made.

Children are still taught about the 'great Saxon invasion and conquest' and of how the new settlers unleashed a torrent of agricultural expansion. This is unfortunate, since archaeology reveals no such sudden conquest and testifies instead to countrysides which were not experiencing colonisation but sliding further and further into decay. With the failure of whatever attempts there may have been to preserve the unity of the Roman province or to sustain York as a great regional capital, the British became divided under the leadership of provincial potentates. A new and relatively durable 'kingdom' developed in Elmet, the lands around Leeds, and Leeds (*Loidis*) may have arisen as its capital (or else *Loidis* could have been a territorial kingdom within Elmet). The names 'Sherburn in Elmet' and 'Barwick in Elmet' tell us a little about its extent. Other more ephemeral British kingdoms may have crystallised in the Dales to the north, and there is thought to have been one based on Craven.

While the evidence of British life is scarce, so too is that of their presumed conquerors. Most Britons would probably have regarded the Saxons as uncultivated ruffians and pirates, and there is some evidence that in the closing decades of the empire and in its aftermath, Saxons were hired as mercenaries to fight their countrymen and other barbarians. Yorkshire has its own Arthurian legends, one claiming that the mythical king and his knights still slumber in the rock beneath Richmond Castle. Even so, there is little to suggest that, during the Arthurian period of the fifth century, the Saxons in Yorkshire were strong or numerous enough to challenge the British polities. As the Roman towns sank into decay, the paid-off or unpaid members of their mercenary garrisons could have been quietly assimilated into the rural societies. On ailing estates, small groups of Saxon settlers might have been encouraged to take over the more run-down lands and, in some places where the power structures had fractured, they might have grabbed such lands for themselves. A trickle of immigration into the East Riding by settlers from the southern Danish and German Atlantic seaboards will gradually have strengthened the Saxon presence in the east of Yorkshire.

A few Saxons settled amongst the decaying grandeur of Catterick at some time in the fifth century. The sunken floor of one of their modest dwellings – a simple hut with a scooped-out floor space – was dug by people who cut its footings through a Roman pavement. At Aldborough, the old standards of urban life were collapsing, and burials within the walled area were now tolerated. Stripped of its Roman garrison at the end of the fourth century, York crumbled, with the marked decline in urbal life coming at about the same time as the garrison was lost, though decay was already apparent by this time. Ottaway writes that: 'In the second half of the 4th century such evidence as there is suggests a gradual decline of building standards and the urban order . . . The street at Rougier Street had dark loam accumulating over it and rubble dumped on it by the mid 4th century indicating that it too was out of use at this time. Pits were found cut into the main road to the south-west at Tanner Row . . .'. York passed out of British control around 600, although the columns of the building that housed the Roman headquarters were still seen standing four centuries after its desertion by the legions. Handmade pottery of the ninth century has been found here, implying that the building was roofed or re-roofed during the Dark Ages. Meanwhile, the sewerage system, which had fallen into neglect in the closing decades of Roman rule, was never repaired, and dust, refuse and the litter of rotting buildings accumulated above the streets and house plots. When the Anglo-Saxons came to York, they probably found that the British people called their decayed capital *Eburac or Evoroc*. They tried to twist their Germanic tongues around the name and it came out *Eofor-wic*, which meant 'boar farm'. When the Vikings took over the town, it became *Jorvik*, though the *vik* or 'inlet of the sea' ending was hardly appropriate. The modern pronunciation was more or less established by early medieval times.

York re-emerged as a great trading capital under Viking leadership in the ninth or tenth centuries, but its Anglian predecessor proved to be elusive, even though the eighth-century poetry of Alcuin of York presented it as a flourishing and cosmopolitan commercial focus. It was eventually located in the mid-1980s, when excavations established that Anglian York lay well to the south-east of the old fortress and *colonia*, to the east of the junction of the Ouse and the Foss, around the cattle market. Excavations beneath the floors of an abandoned glass factory explored the remains of a Gilbertine house; beneath these were found rubbish pits dating from the eighth century. Seven or eight timber-framed Anglian halls were discovered, with associated gravel roads, while discoveries of continental pottery, craftwork, textile manufacture and metalworking established the commercial nature of Anglian York. Subsequent finds suggested a substantial settlement, and Kemp writes that: 'Extrapolating from this, it looks as if the Anglian settlement might have covered as much as 25ha, and far from having been a huddle of merchants on the side of the river, seems to have extended over an area comparable to Quentovic (in France), Ipswich, Londonwic or perhaps even Hamwic.' It seems that the occupation of Anglian York ended when the Vikings gained control and developed the site at Coppergate in the old Roman area.

The date of York's resurgence from the decay that followed the withdrawal of the Roman army is uncertain. One suggestive but puzzling monument is the famous Anglian Tower, a barrel-vaulted structure which closes a gap in the Roman defences, having been inserted at a point where the Roman wall had suffered a structural weakness, perhaps caused by the collapse of a postern or external tower. It was regarded by Radley as being a Saxon feature and his ideas found much support, hence its name. Others, like Buckland, have located the origins of the 'Anglian' Tower in the late Roman period. In any event, by the late seventh century, York had reasserted its presence as a dominating centre in the Yorkshire landscape. It was during this century that English-speaking peoples achieved their mastery over Yorkshire. Coming as poor settlers or soldiers with nothing to lose, the Anglo-Saxons must have proved more

The Anglian Tower at York was discovered when a trench was dug in 1839. Its construction is different from that of any other part of the Roman defences and it is presumed to be a tower, once much taller, which was built to plug a gap in the Roman walls in the period AD 600–700

resourceful and adaptable than the more sophisticated but demoralised indigenous populations. Numerically, the newcomers would have been greatly inferior to the British, but, perhaps because they provided military leadership or managed to gain control of the leading estate centres, they were often able to assert a political mastery. Gradually, and by means that are still unknown, their language gained ascendancy over the British dialects. (Celtic names were retained for many geographical features – rivers like the Nidd, Derwent and Wharfe, hills like Pen-y-Ghent, and districts like Craven. Interestingly, surviving Celtic-based dialect words in Nidderdale show a fairly equal balance between those derived from the Gaelic of Ireland on the one hand, and the Celtic dialects of Wales, Cornwall and, presumably, England on the other.)

The nationalist movement of the nineteenth century and the traditional English paranoia about invasion have tended to distort interpretations of events and issues in post-Roman Britain. The period is often regarded in terms of a massive confrontation between nations, involving Roman-ised Celts and Britons on the one hand, and invading Anglo-Saxon hordes on the other. In fact, most Dark Age armies were modest in size and the political destinies of large territories could be determined by small numbers of warriors – as, meanwhile, the peasant masses continued to work the land, perhaps somewhat indifferent to the struggles between the established and aspiring élites. The traditional scenarios in which the native British are exterminated or driven into the fastnesses of the west receive little if any support from informed modern opinion. From major 'Anglo-Saxon' excavations as far apart as Yeavering in the north-east to Sutton Hoo in the south of East Anglia, the picture that emerges is one of an essentially British population with a veneer of Saxon leadership. The same situation probably applied in Yorkshire.

In 547, King Ida of Bernicia established his centre of government at Bamburgh, 'the fort of Bedda' (his queen), on a sea-flanked volcanic out-crop to the north of the region. At first, the Tees formed the boundary between the vestigial king-doms of Bernicia to the north, and Deira, covering the East Riding, to the south, but the merging of these two territories produced the powerful and expansionate kingdom of Northumbria, with its important place in the ancestry of Yorkshire. Towards the end of the sixth century, the people under Saxon influence began to spill outwards from their lands in the old East Riding to link with the Saxons living further north. Legends preserved in the early Celtic literature tell of a great Saxon victory at Catterick around AD 600 which caused the death of several British leaders. This would have paved the way for the union of the English-dominated kingdom of Deira with Bernicia to form Aethelfrith's new kingdom of Northumbria. The death-knell for British inde-pendence sounded in 617, when the army of King Edwin of Northumbria (616–32) overran the Celtic bastion of Elmet, and Ceretic, son of Guallac, the last British king to reign in Yorkshire, was expelled.

There are no positively identified monuments to these struggles, although sets of frontier works do seem to record the insecurities of political life in the Dark Ages. There are puzzling banks to the east of Leeds in the Aberford area; South Dyke might have been built to protect *Loidis* against the westward advance of the English, while the Becca banks might have been built a little later, perhaps to protect Northumbria against an intrusion from the expanding Midlands kingdom of Mercia, although it has also been interpreted as a Brigantian defence work to check the Roman advance. While the South Dyke faces north, the Becca banks face south, and the remains tell of a formidable earthwork, its crest standing about six metres above the foot of its facing ditch it may also have carried a timber palisade. About four miles (6.4 km) to the east of Leeds is Grim's Ditch, formerly mistaken for a Roman road, although the best-preserved sections of the bank still stand almost 2.4 metres high. Along with South Dyke and Rudgate Dyke to the west of Tadcaster, Grim's Ditch is probably part of a defensive complex of banks and ditches built to resist the Saxon advance on Elmet.

While the political map of the North was being forged, the gradual re-emergence of Christianity offered at least the hope of a return to more civilised values. Official attitudes to Christianity had fluctuated during the period of the Roman empire, and it is not easy to assess the appeal of the religion amongst the very different sections of society in Britain. However, it is known that a Bishop of York attended a church council in France, at Arles, in 314. After the Roman collapse, the rudiments of Christian worship were pre-served in the wild rocky lands of the west of Britain, where monasteries were established, and these formed the bases from which Christian conversion was returned to some parts of England. One is left to wonder whether, through all the decay and turmoil of the fifth and sixth centuries, the candle of Christian worship was kept burning in Yorkshire. Quite probably it was, and *The Life of St Wilfrid*, which was written about 715, mentions consecrated places which

Becca banks, still an impressive bank-and-ditch earthwork, part of a puzzling complex of frontier works in the vicinity of Aberford, near Leeds. It may, like others in the complex, have been part of the defences of the British kingdom of Elmet, but it was perhaps a frontier work of Northumbria, built to resist penetration by Mercians

were abandoned by the British clergy as the Saxons invaded Elmet. Clues come from place-names which derive from the vulgar Latin *eclesia*, 'a church', which would be pronounced 'eglesia' by the British and which emerges in forms like 'Eccles' in later parlance. Such words certainly seem to denote places which supported Christian congregations before being occupied by the Saxons, who used *cirice* (preserved as 'church' or 'kirk') to describe their church buildings. Faull and Smith (1980) write:

It seems likely that the English borrowed the term *eccles* to describe a British ecclesiastical establishment in contrast with an English *cirice* and that, once the British churches ceased to function, the term became superfluous and disappeared from the Anglo-Saxon vocabulary. The alternative possibility must also be considered that the term had already been used by the British to describe an earlier church, which need not still have been in existence when the English arrived, and that it was the British

place-name formed from **egles* which the English borrowed, rather than the element itself. Whichever was the case, the occurrence of the element *eccles* in a place-name should indicate that at some date there was a British church in the area.

If we look at the distribution of 'Eccles-type' names in Britain, then the eye is instantly drawn to a cluster of no less than nine such names which coincide with the old kingdom of Elmet: Ecclesfield, Eccleshill, Ecclesall, and so on. If the reasoning is correct, then nobody in Dark Age Elmet would have lived very far from a place of

Christian worship. Three of the 'Eccles' names in Elmet were field names, raising the question of whether these fields had long ago been the sites of British churches. In an attempt to solve such an apparently insoluble riddle, Faull and Smith carried out phosphate analyses of the soil in these fields, reasoning that bones from a British cemetery might still be evident as high phosphate levels in the soil. The results were unremarkable in two cases, but Ecclesfield, Stanbury, near Haworth, did produce the high phosphate recordings to support the field-name evidence for a British church site.

Whether any form of diocesan organisation survived, we do not know, but the years that followed saw vigorous debate as to which form of Christianity would triumph in the reconverted lands – the Celtic monastic or the Roman episcopal – and Yorkshire lay in the marchlands of the confrontation. In 597, Augustine's mission from Rome landed in Kent. Within three decades the reconversion to Christianity had penetrated the northern lands and Augustine's disciple, Paulinus, was installed as the new, though not the first, Bishop of York. Kent was fairly safely in the Christian fold and Edwin of Northumbria sought a marriage alliance with the Kentish dynasty. When King Eadbald replied that his sister could only marry a Christian, Edwin adopted the new faith and in 627 Northumbria (re-)entered the Christian world.

Meanwhile, Celtic missionaries were, belatedly, becoming involved in converting the English peoples. Columba had established monasticism on Iona in 563, and in about 634 an offshoot from this community came to Lindisfarne, an island base for the conversion of Northumbria. In 633, however, Paulinus had been obliged to flee from York following the death of Edwin in a battle with the Welsh and Mercians at Hatfield in the previous year. Cadwallon and his pagan Mercian allies were soon ejected from Northumbria, while Oswald, the victor, as the son of Aethelfrith, a convert of the monks of Iona, was sympathetic to the Celtic church. It was at Oswald's invitation that Aidan brought a company of monks to Lindisfarne. James the Deacon, although a friend of Paulinus, continued missionary work in the north, having retreated to the relative safety of Catterick, the initiative now passed to the Celtic missionaries, and in about 635 they began their work on the Northumbrian mainland. So now two versions of the one religion competed in the northern lands, the one austere and monastic, the other organised on episcopal lines. To resolve the differences, King Oswy of Northumbria convened a synod at Whitby in 663, with Colman, Bishop of Lindisfarne, speaking for the Celtic cause, and Wilfrid, Abbot of Ripon, arguing the Roman case, the latter appears to have been the most persuasive. Although the dispute centred on methods of calculating the date of Easter, the real issue was one of geostrategic importance; defeat at Whitby represented a crushing rejection for Celtic traditions in Yorkshire.

Very little is known about the first generation of Christian churches in Yorkshire, but they were probably almost all of them simple wooden structures. Bede records that, when Edwin was baptised at York on Easter Day in 627, a new church, dedicated to St Peter the Apostle, was hastily built of timber to accommodate the ceremony. The conventional version of the subsequent events runs as follows. Some time after being baptised in York, Edwin probably replaced his baptismal church with one of stone. This was burned in 741 and a Saxon cathedral was built on the site. It in turn was burned by the Normans, although a portion of its eighth-century walling endures in the crypt of the medieval cathedral. A different interpretation, focusing on the notion that the south-west bank of the Ouse supported a major urban nucleus in Roman times, has been suggested by Palliser, who wrote:

It is quite possible that Paulinus, coming to York in 625 as the first Anglo-Saxon bishop, would have taken over ... [any surviving Roman church there] if it were still standing. Traditionally, of course, his episcopal church is equated with the oratory and church of St Peter in which, according to Bede, he baptized King Edwin in 627, and with the site or neighbourhood of the present Minster. However, Edwin's church has not been found; and furthermore, nowhere does Bede explicitly identify it with Paulinus's cathedral. Even if the Anglo-Scandinavian cathedral stood on or near the site of the present Minster (as the burial ground found beneath the south transept suggests), there could well have been a situation in which, in the 7th and 8th centuries, a cathedral on the south-west bank co-existed with a private royal chapel on the north-east, or two or more major churches might have been in use simultaneously. No one has yet located the *monasterium* burned in 741.

Following Edwin's conversion, York became the effective capital of the northernmost of Britain's two archbishoprics, although, because of the unrest in Northumbria, it was not raised to metropolitan status until 735. Around the time of Edwin's baptism, a new church was built in Elmet. The Christian Church offered more than just spiritual guidance, for it embraced educated, literate men who could record the financial and legal affairs of a kingdom. It was probably for such reasons that the early Christian foundations were frequently established on royal estates; they were not parish churches, but minsters, staffed by a body of churchmen who took the gospel out to

The doorway of Ledsham church

Ledsham church, near Leeds, a very early foundation which may have originated as a royal monastery

the communities of the surrounding townships. Parish or 'field' churches came later, and most parishes gained their ecclesiastical status in the last centuries of Saxon rule.

One of the most important early churches in the region is at Ledsham, near Leeds. It seems to preserve the old *Loidis* district name and could have originated as a minster and monastic centre, perhaps a royal monastery. The seventh-century nave is almost complete, having survived a Victorian restoration; the base of the tower and the south porch are of the same date. It must take one back almost to the time when Christianity was restored as the state religion in Northumbrian lands. The doorway into the tower is decorated with carving which is not authentic Saxon work and which probably dates from a restoration of the church by the architect, Henry Curzon, in 1871. Another important early church site seems to have been at Sherburn in Elmet, where the church, with its largely late Norman fabric, stands in an ancient but undated earthwork enclosure. A Saxon cross on a later shaft stands in the churchyard. It is thought that such circular or oval churchyards can indicate pre-Saxon church sites,

The Church at Sherburn in Elmet stands in a vast earthwork enclosure and appears to be derived from a very early focus of Christian worship

in which case the oval churchyard at Bramham, near Tadcaster, could denote a very ancient site.

Yorkshire boasts numerous Dark Age crosses, but it is seldom certain of these marked preaching places, whether they were churchyard crosses, whether they were erected as tombstones, or whether they mark the boundaries of church lands. Nor is it always easy to identify the original location of a cross, since some were removed and re-erected, often in churchyards. We know that at least some of the crosses marked preaching places, for the ninth-century *Life of Willibald* informs us that, on many estates, there was no church, only a cross raised on high to mark the place where daily prayers were said. But the crosses also seem to have been versatile in their use, and Bede tells of how Oswald set up a wooden cross before advancing his troops at the battle of *Hefenfelth*. Three Saxon crosses, one of about 800 and two of about 850, stood together in the churchyard of All Saints, Ilkley, which was fitted into a corner of the Roman fort. Two are

quite short, one of them having acquired the head from another monument, but the tallest is a fine example with the signs of the four evangelists on one face and Christ above beasts on the other. The neighbouring town of Otley has less celebrated crosses. The estate here was granted to Wilfrid in or before 678, and thus became a possession of the Archbishops of York. A monastery was founded at Otley in the following century, while the archbishop had a residence at Addingham. Two Anglian crosses stand at Otley, the Angel (or Otley I) cross being dated to the late eighth century, and the Otley II cross, with its mythical beasts, probably belonging to the period before Archbishop Wulfthere fled to Addingham when the Vikings captured York in 876. According to Wood, the Otley crosses may have served primarily to emphasise the fact that the estate at Otley belonged to the archbishops, providing a focus for pastoral activity. The parish church of St Peter in Leeds has a superb cross in its churchyard, a late Saxon creation standing over 3.4 metres tall. The finest cross in the east of Yorkshire is the one preserved in Nunburnholme church, dating from about 950. Viking pagan mythology was still fresh in the public mind at this time and various figures from the myths are given a place on this Christian

symbol, the Virgin and Child sharing the cross with Sigurd devouring the dragon's heart. In the churchyard at Masham is a badly eroded ninth-century cross with beasts set in arcades on its shaft, while further north, at Croft, is a finer carving. An interesting composite monument can be seen at Holy Trinity, Stonegrave, near Malton: a wheel-headed tenth-century cross is set on a thirteenth-century grave-slab, which in turn stands

The enormous oval churchyard at Bramham could denote a very ancient church site

on an Anglo-Danish slab which is decorated with a man shooting a dragon. Many crosses must have had chequered histories, suffering from the attentions of frost, iconoclasts, masons short of building stones, and farmers needing gateposts. Some of Yorkshire's best examples now survive only in fragmentary forms. To the east of Ripon, the churches at Cundall and Aldborough share fragments of the same cross, while important collections of incomplete crosses can be seen at All Saints, Dewsbury, and several other churches.

Christians and non-Christians alike are moved by the thought that worship may have continued in one place for a thousand or more years. The first churches in Yorkshire may mostly have been built of timber, but some churches of stone will have been built in the seventh century. The Saxon period of church building in Yorkshire spanned more than four centuries, so that any church site could have experienced several phases of Saxon rebuilding, repair and improvement. This point is apparent to visitors to the church of St Gregory at Kirkdale, near Helmsley, where the famous Saxon sundial is inscribed: 'Orm, son of Gamal, acquired the church of St Gregory when it was tumbled and ruined, and had it rebuilt from the ground in honour of Christ and St Gregory, in the days of Edward the King [Edward the Confessor, 1042–66] and Tosti the Earl' (Tostig, brother of

This Saxon cross, probably of the tenth century and signifying a site of Saxon worship, was discovered in the foundations of the chancel of the church at Brompton-in-Allertonshire

King Harold). The church is on the flood plain of the Hodge Beck and was probably preceded by a Saxon monastery and its estate. We know that the Gamal who is mentioned was one of the unfortunates murdered by Tostig in 1065, prior to the earl's timely banishment. Those with a good eye for architecture will also recognise the evidence of Saxon rebuilding at All Saints, Kirby Hill, near

(Left) *A beautifully decorated cross of the late seventh or early eighth century in Easby church, near Richmond. A reconstruction can still be seen there. The original is now in the Victoria and Albert Museum*

(Below) *The Saxon sundial at St Gregory's, Kirkdale, is the best of its kind in existence. The daylight period is divided into four 'tides', which are then subdivided. The inscription tells how 'Orm, Gamal's son bought St Gregory's Minster when it was all broken down and fallen, and he let it be made anew from the ground to Christ and St Gregory in the days of Edward the King and Tosti the Earl. And Haward wrought and Brand Priests' is the inscription on the sundial panel. The text is in Northumbrian English and dates to the period 1055–65*

Boroughbridge, where sculpture and masonry have been reused and incorporated in the walls. Similarly, at the church of St Peter's at Hackness, near Scarborough, the chancel arch is regarded as eighth-century work, probably reused during the building of the nave in the following century. However, it is known that Abbess Hilda of Whitby built a monastery at Hackness in 680, and so the sequence of Saxon churches here must have succeeded the monastic foundation.

Stone appeared in church buildings long before it was used for domestic architecture. Cut stone was sometimes pillaged from nearby Roman ruins, like the stone displayed in the base of the tower of the attractive church at Little Ouseburn. Faull has suggested that York might have expressed its leadership in ecclesiastical affairs by exporting cut gritstone for use in church building, the stone perhaps travelling as ballast along the river trading network. To see an almost perfectly preserved stone church of the late eighth century one must step just outside the region to Escomb, near Bishop Auckland, where the stones were gathered from the ruined Roman fort at Binchester. Yorkshire itself is rich in Saxon churches, Ledsham being an exceptionally early survivor; the church at Kirk Hammerton has a nave and chancel dated by some to the seventh or eighth centuries, though others would regard the church as a very late Saxon building with an early or 'timeless' form. The tower is also Saxon, belonging to the tenth century. At Skipwith, near Selby, the lower part of the eleventh-century tower is formed from the porch of an early Saxon church.

The most celebrated example of eighth-century church architecture in the region is the crypt beneath Ripon Cathedral. St Wilfrid, who debated in front of King Oswy at Whitby, established his monastery at Ripon and was buried to the south of its altar. He became Abbot of Ripon in about 660 and Bishop of York in 669; during his term of office, he established churches with crypts at both Hexham and Ripon. The church at Ripon was described as a stone building with columns and had side-aisles. The crypt is the only surviving portion of the monastery, which was burned by the Danes in the ninth century and again by Eadred, the first King of England, who raided a rebellious Northumbria in 948. Subsequently, it was rebuilt as a church, destroyed by the Norman army in 1080, vandalised by the Scots in the fourteenth century, repaired, devastated by Cromwell's supporters, and eventually adopted as a new cathedral in 1836. In 1986 a new facet of Ripon's long history emerged during exploratory excavation at the glacier-dumped mound of gravel known as Ailcey Hill, which lies 200 metres to the east of the cathedral. A number of iron-bound coffins of the Viking era of the ninth and tenth centuries were unearthed, and the existence of an enormous contemporary cemetery here was confirmed. Evidently, Ripon was as a substantial and influential centre during this poorly understood and unsettled period.

Work dating to the ninth century can be seen at All Hallows, Bardsey, near Leeds, where the lower section of the tower was originally the porch, with the remainder of the tower being raised in the tenth century. The south doorway of All Saints, Kirby Hill, Boroughbridge, with its interlace and vine-scroll decoration, is also of the tenth century. A number of Yorkshire churches contain work dating from the last century or so of the Saxon period. One of England's most attractive little churches is at Wharram le Street, standing a little apart from the village. There is Saxon masonry in the nave and tower and four original windows in the belfry. The tower at Monk Fryston, near Leeds, belongs to the final phase of Saxon church building; by the time that the level of the belfry was reached, the Normans had arrived in England. Other late Saxon towers can be seen at Hovingham, near Malton, and St Mary Bishophill Junior, York.

Much is still to be learned about the early history of the Church in Yorkshire. Several minster sites are as yet unrecognised and much is still uncertain. For example, it is suspected that the vast manor of Wakefield may have been a large royal estate in Saxon times, served by a minster church at Dewsbury. The sites of several small monastic houses that never recovered from the Viking onslaught also await discovery. Saxon monasteries existed at Ripon and Sherburn and at Otley. The Venerable Bede mentions a seventh-century monastery of Abbot Thrythwulf in the forest of Elmet at the royal estate of *Campodunum*, and there are grounds for identifying *Campodunum*

Stones from a ruined Roman building were used in the base of the Saxon tower at Little Ouseburn church

as Leeds. Excavations carried out in the mid-1980s at the newly discovered site of Anglian York

confirmed that churches of timber were still being constructed at the end of the Saxon era. The example found was provisionally dated to the eleventh century and had a floor of clay. About 40 bodies were found in the associated churchyard; it was assumed to be the parish church of St

Andrew, mentioned in Domesday Book. Soon after the Norman conquest, the timber church was superseded by a larger church of stone. At this time, according to Domesday, England contained 2,700 local churches; the true figure would have been higher, however, as numerous examples are known to have been overlooked.

Had any citizen of Roman Britain been able to rise from the dead and explore late Saxon Yorkshire, he or she would be shocked by the relative neglect of the countryside and by the still-evident signs of decay. Such a person would see the modest attempts to reproduce classical architecture in the churches that were springing up, but would see no buildings as grand as those of the greater Roman towns or fortresses, and nothing to compare with the refined domestic life of the villa. Nevertheless, a peasant from the fifth or sixth century would have been gratified to witness the extent of the recovery and, perhaps, rather surprised to see that all the most impressive architectural and intellectual achievements were actually the accomplishments of the Christian

Church. It is also worth wondering how British people living in Yorkshire reacted to the tumultuous political and ecclesiastical changes which occurred at the start of the seventh century. At the end of the sixth century, the legends of the achievements of Roman Britain would still have been alive, sources both of pride and regret. Yet, virtually within the space of a single generation there was the British defeat at Catterick; the fall of the kingdom of Elmet; absorption into Northumbria; reconversion; the reactivation of the

The church at Hackness stands near or on the site of a Saxon nunnery which was destroyed by Viking raiders in 867–9. The Saxon chancel arch may have been built in the eighth or ninth century as part of the monastery. The monastery was refounded in 1086–96, but was probably abandoned by its community in favour of Whitby in about 1100. Norman (right) and Gothic (left) columns flank the aisles of the historic church, and the Hackness cross (see p.110) can just be glimpsed beyond the Norman pier

diocese of York; and the baptism of Edwin in 627. Later, there would have been a few British people alive in Yorkshire whose memories could span the Catterick defeat of around 600 and the effective vanquishing of the Celtic Church at Whitby in 663. But even people as chastened as these would be surprised if they could return and see how even comparatively recent history books devalued the Celtic contribution to Yorkshire's cultural landscapes.

REFERENCES

C. J. Arnold, *Roman Britain to Saxon England* (London: 1984).

P. C. Buckland, 'Archaeology and environment in York', *Journal of Archaeological Science*, 1 (1974), pp. 303–16,

P. C. Buckland, 'The "Anglian Tower" and the use of Jurassic limestone in York', in P. V. Addyman and V. E. Black (eds), *Archaeological Papers from York Presented to M. W. Barley* (York: 1984), pp. 51–7.

K. Cameron, 'Eccles in English place-names', in M. W. Barley and R. P. C. Hanson (eds), *Christianity in Britain 300–700: Papers Presented to the Conference on Christianity in Roman and Sub-Roman Britain Held at the University of Nottingham 17–20 April, 1967* (Leicester: 1968), pp. 87–92.

W. G. Collingwood, *Northumbrian Crosses of the Pre-Norman Age* (1927).

M. L. Faull, 'Britons and Angles in Yorkshire', *Studium: The Journal of the Sydney Medieval & Renaissance Group*, 6 (1974), pp. 1–24.

M. L. Faull, 'British survival in Anglo-Saxon Northumbria', in L. Laing (ed.), *Studies in Celtic Survival* (Oxford: 1977), pp. 1–56.

M. L. Faull, 'Place-names and past landscapes', *English Place Name Society Journal*, 11 (1978–9), pp. 24–46.

M. L. Faull, 'The pre-conquest ecclesiastical pattern', in M. L. Faull and S. A. Moorhouse (eds), *West Yorkshire: An Archaeological Survey to AD 1500*, 4 vols (Wakefield: 1981), i, pp. 210–23,

M. L. Faull, 'Roman and Anglo-Saxon settlement patterns in Yorkshire: a computer-generated analysis' *Landscape History*, 5 (1983), pp. 21–40.

M. L. Faull, 'The decoration of the south doorway of Ledsham church tower', *Journal of the British Archaeological Association* 139 (1986), pp. 143–6.

M. L. Faull and R. T. Smith, 'Phosphate analysis and three possible Dark Age ecclesiastical sites in Yorkshire', *Landscape History*, 2 (1980), pp. 21–38.

M. Gelling, *Signposts to the Past: Place-Names and the History of England* (London: 1978).

J. Grenville and P. Rahtz, 'Archaeology in Kirkdale', Supplement, *The Rydale Historian*, 18 (1996–7).

N. Higham, *Rome, Britain and the Anglo-Saxons* (London: 1992).

G. F. Jensen, 'Place-names and settlement history: a review with a select bibliography of works mostly published since 1960', *Northern History*, 13 (1977), pp 1–26.

R. Kemp, 'Anglian York – the missing link', *Current Archaeology*, 104 (1987), pp. 259–63.

L. and J. Laing, *A Guide to Dark Age Remains in Britain* (London: 1979).

P. Ottaway, '*Colonia Eburacensis*: a review of recent work', in P. V. Addyman and V. E. Black (eds), *Archaeological Papers from York Presented to M. W. Barley* (York: 1984), pp. 28–33.

D. M. Palliser, 'York's West Bank: medieval suburb or urban nucleus?', in P. V. Addyman and V. E. Black (eds), *Archaeological Papers from York Presented to M. W. Barley* (York: 1984), pp. 101–8.

J. Radley, 'Excavations in the defences of the City of York: an early medieval stone tower and the successive earth ramparts', *Yorkshire Archaeological Journal*, 44 (1972), pp. 38–64.

A. H. Smith, *Place-Names of the West Riding of Yorkshire*, 8 vols (Cambridge: 1962).

C. Thomas, *Christianity in Roman Britain to AD 500* (London: 1981).

N. Wood, 'Anglo-Saxon Otley: an archiepiscopal estate and its crosses in a Northumbrian context', *Northern History*, 23 (1987) pp. 20–38.

CHAPTER EIGHT

English and Danish Settlement

MUCH HAS BEEN WRITTEN ABOUT the surviving architecture of the Saxon centuries, far less about the landscapes in which the buildings stood. There is no doubt that, at the end of the Roman period, the countrysides of Yorkshire fell into decay. Taking parts of the North York Moors (where the fossilised pollen evidence has survived quite well) as an example, it appears that areas which existed as open farmland in the Iron Age and Roman periods experienced a regeneration of woodland during the Dark Ages which continued until the late Saxon or Viking period, with a fresh, large-scale assault on the woodland taking place in Viking or Norman times. Even in the later centuries of the Middle Ages, however, peasants were still felling woodland that had become established during the retreat of settlement following the Roman withdrawal. One must ask whether these deeply troubled times witnessed the decay of all the established patterns and institutions of landownership, or whether the rural framework that had existed in Roman Yorkshire still exerted an influence on the countrysides of Viking and Norman times.

Some clues may be found by looking at parishes. There were no parishes as such during the first centuries of Saxon Christianity in Yorkshire, and teaching and worship were organised through the system of mother or minster churches. It was only in later Saxon times, when local landowners were building new churches for their own use and for use by their tenants, that parish churches were established, the parish being the local territory which supported and was served by a church. (The personal link between a church and the local notable who created it is very occasionally preserved. This occurs at Aldbrough, near Hornsea, where the church contains a Saxon sundial inscribed with a dedication: 'Ulf, who ordered this church to be built for his own and Gunware's souls'.) With the rapid spread of 'field' churches of this type in the later Saxon centuries, ecclesiastical parishes were clearly needed. They did not need to be invented, however: they already existed in the form of rural estates. These existing estates were likely to be very old indeed, and it is sometimes suggested that they have prehistoric origins. The most important type of landholding in Roman and Celtic Britain seems to have been what is now often called a 'multiple estate'. This was a substantial area of land embracing many different farms and settlements and types of countryside. It could have various detached portions; there seems to have been distinct specialisation within the multiple estate. Some localities might produce grain, some would raise horses or oxen, and so on – so that each estate was a self-contained entity which could also engage in commercial production. In the course of the Dark Ages, these large estates seem often to have become fragmented, yet their essential integrity often survived. Taking an example from Nidderdale, Jones has described how, at the time of Domesday, in 1086, the area between the Wharfe and the Ure existed as a large division or 'wapentake' known as 'Burghshire' (*Borgescire*). Within this wapentake were two royal multiple estates, one focusing on Aldborough and one on Knaresborough. He believes that Burghshire was the direct descendant of the tribal district of the Brigantes which had *Isurium Brigantum* or Aldborough as its capital in Roman times. (*Burh* is a Saxon word for a fortified place, and would describe the ruined walls of Aldborough very well.) In the course of the Dark Ages, the Roman tribal territory must have fragmented to produce

Nidderdale, from the churchyard at Middlesmoor. The name combines the Celtic river name of 'Nidd' ('brilliant water') with the genitive of the Scandinavian word for a valley. This is one of the many hybrid place-names in Yorkshire

the two royal multiple estates described in Domesday Book, while smaller fragments of land passed to lesser lay and ecclesiastical owners. Here one can add that memories of the Burghshire territory lingered on; although the land became divided between many holders, the name 'Riponshire' was used to describe the land between the Ure and the Nidd in medieval times. It seems that a system of partible inheritance was normal in old Northumbria, with estates being divided between male heirs; naturally, this would lead to the division of great estates.

In a recent survey, Hadley notes that:

> many have rejected the model chiefly on the grounds that the evidence used by Jones is very late, and they are unable to accept that the social structure and forms of exploitation he describes could ever have characterized the medieval period. Nonetheless, although many studies would seek to distance themselves from that particular model, many have accepted the basic principle that early-medieval society was commonly organised in multi-vill estates, and numerous studies have sought to identify a network of such estates through the use of documentary, cartographic and place-name evidence. These have come to be known as either 'multiple' or 'federative' estates or by their local name of shire, soke, lathe and so on, and one historian has commented that it is really only the choice of local terminology that varies in the pattern of local organization that has been identified.

She concludes:

> Much recent research indicates that the provision of tribute to superior powers was effected through large territories in the early medieval period and collected through administrative centres, often royal *tuns*. A number of the large Domesday sokes owed food renders and other provisions to the king, which appear to represent a remnant of the ancient tribute system. Furthermore, it is not difficult to envisage that at an early date such territories contained inhabitants and land of varying status, and as such the Domesday sokes reflect an early type

of organization. However, this does not prove that the precise distribution of sokes revealed in the tenth and eleventh century is equally ancient.

In Swaledale, a project on ancient land boundaries directed by Fleming and Laurie found that ancient linear earthworks provided clues to the partitioning of the dale. Fleming wrote: 'The linear earthworks suggest the existence of a hitherto unrecognised early – post-Roman polity in Upper Swaledale and Arkengarthdale – a polity which might be held to correspond, at least roughly, with the large medieval parish of Grinton and the manor of Healaugh'; and: 'It must be a matter for debate whether the Grinton-Fremington dykes were built by a "British" polity – Erechwydd? – or by an "Anglian" polity – by the people of Swar . . . in the fifth, sixth and perhaps early seventh centuries . . . In any case they must have been constructed mainly by the descendants of the (evidently substantial) Romano-British population of Upper Swaledale.'

It seems highly probable that many of the ecclesiastical parishes of Saxon Britain were rural estates which represented fragments or components of the substantial landholdings of Roman or earlier times – and, as such, they would often be very old territories indeed. However, popular ideas about parishes, villages and Domesday Book tend to be confused and confusing. Regarding the latter, the most important thing to remember is that Domesday Book was not intended as a description of villages and countryside which might be of use to future historians, but as a list of taxable lands, their contents, chattels and holders. As a record of assets taxable by the king, it was indifferent to the distribution of settlement within a spatial unit of taxation: it does not list the taxable assets of a particular *village*, but of a *small estate*; no village is necessarily implied. Thus, Birstwith in Nidderdale is mentioned in Domesday, but had no existence as a village (rather than as a land unit) until the nineteenth century. Turning to the parish, popular perceptions are often derived from a crude model of Saxon England which is based on an idealised example from the English Midlands, though the model does not even fit the Midlands very well. We are taught to think in terms of the 'big village', surrounded by village lands which equate to the parish. In Yorkshire, where land can be quite poor and the population dispersed in farmsteads and hamlets across the face of a large parish or old estate, the township rather than the parish was normally the fundamental unit of community. For example, the one parish of Halifax contained some 22 townships and, in the thirteenth century, each of these townships had 30–40 minor settlements within it. The former parish of Lythe in the North York Moors spanned 9 townships and included 11 villages.

A more reliable characterisation of the northern landscape in early medieval times would envisage a countryside which is divided into many small rural cells, each cell supporting a little community. These cells can be called 'townships' or 'vills' and, within each vill, a small peasant assembly could organise local affairs. A manorial estate or an ecclesiastical parish (the two often coinciding) was likely to embrace several or many vills and, in due course, the vills, as components usually of more extensive estates, were likely to have their relevant contents listed in Domesday Book. Vill, parish and estate could coincide in the richer countrysides of southern England, but in the North, particularly in the less productive countrysides, a lord of any substance would need to control several vills if he was to maintain a standard of living appropriate to his station. The antiquity of the vills is not known. They can sometimes be traced back to the early stages of the Christian era before they recede from view, but, as with the much larger multiple estates, they could be very old indeed. Faull notes that, in Yorkshire at least, the vills were well established by the time of Domesday Book: 'It is possible that the vill as a unit may be of considerable antiquity and that the institution of units belonging to groups who exploit the entire landscape as a community and practise communal agriculture may go back to before the Anglo-Saxon settlements.' As the centuries rolled by, a formerly useful vill or township might lose its value to the local community and be forgotten. One of its farmsteads might form a nucleus for settlement, and a hamlet might develop around it. With further growth, the hamlet might gain its own chapel of ease, and in time the chapel could become the church of a new parish. As a result of such changes, the rural landscape today is perceived in terms of parishes, villages, hamlets and farmsteads. Township names often linger on, attached to hamlets or to localities, but the township boundaries – and their former importance in defining the bounds of community – may be quite forgotten.

In summary, the countrysides occupied by the Anglians and Vikings were criss-crossed by boundaries of different kinds, many of them very old. Landowners of different grades held estates, some large, some small, most of them probably derived from the splitting-up of ancient multiple estates. Churches dominated the parishes, many of which equated to estates as a result of estate owners providing field churches to serve their families and tenants. Peasant communities inhabited vills, some containing villages and some devoid of them but embracing a scatter of hamlets and farmsteads. Some of the main decisions about peasant responsibilities and estate management were taken at the estate or manorial level, but the day-to-day affairs of peasant life revolved around the vills: were one to ask a Saxon or Viking country dweller where they lived, they would give the name of their vill. A parish or estate might encompass but a single vill if it lay on good lands in, say, the Wolds or on the Magnesian limestone belt; more often, it would comprise four, five or six vills, sometimes many more. The parish of Helmsley encompassed 40,000 acres, and that of Pickering 31,000 acres. Grinton was another gigantic parish, but, as noted above, the parish here was derived from an earlier political territory, and Fleming notes that: 'The entity which became both Healaugh Manor and Grinton parish in the twelfth century was already called Swaledale by this time; in a political rather than a geographical sense, Swaledale was an alternative, and obviously older, name for both parish and manor. Medieval documents in the *Bridlington Cartulary* refer to Grinton church as "the church of Swaledale" . . .'.

Almost everyone was a peasant of one kind or another in Saxon and Viking Yorkshire (even though the term 'peasant' is sometimes rather pedantically reserved for the manorial tenants of the post-conquest manors). Lords, warriors,

A magnificent trio of hogback tombs in the church at Brompton-in-Allertonshire

churchmen, artisans and merchants together amounted to only a small minority, but they all depended for their existence on the labour of the peasant masses. In the late Iron Age, the need for land had been so great that new fields in the Wolds were run across existing cemeteries of tightly packed square barrows; the conditions in Roman Britain demanded even more of the native farmers. From the evidence that survives, this farming seems mainly to have been practised in small rectangular fields. Even in remote and rugged Swaledale, the countryside appears to have been well-settled in Iron Age and Romano-British times, and the linear earthworks of the Dark Ages were built across the field banks of these earlier periods. Subsequently, chaos and decay afflicted the post-Roman countrysides;

given the economic disarray, there was no point in producing for markets which no longer existed or could no longer be reached. Landowners may have fled with the breakdown in law and order, and their peasants could have lost their homes, beasts and crops whenever war-bands stalked the land.

As a consequence of the uncertainty and instability, a patchy sort of countryside is likely to have come into existence: here a cluster of working farms; there thorns and briars advancing across the fields of a community which perished in a plague, beyond a vill where people were struggling against the odds to recover from the depredations of the latest raiding party; and over there a forceful new lord organising his peasants to clear the overgrown land. Some land had been so overworked that it would be barren and in need of fallow. In some places, the loss of livestock would have removed the means of preserving pasture against the spread of gorse or woodland,

A detail of one of the Brompton hogbacks. The stone tomb cover – perhaps symbolising a house – is clasped by muzzled bears. Above the decorative interlace patterns can be seen a scale-like pattern, perhaps representing a shingled roof

while in others the margins of farming must have contracted as depleted communities concentrated their efforts on the choicest lands. The nature of the change from British to Anglian dominance was complicated and various systems of accommodation must have existed. In Upper Swaledale, Fleming thought that:

> An argument could be made for the Cogden estate being a survival of an earlier pattern of land-holding, of the sort represented archaeologically by the long ruined walls on Grinton Moor, which form narrow strips running across the contour and pre-date the Grinton–Fremington dykes ... Grinton and Ellerton may have been recognising a pre-existing territory of 'pre-Anglian' character, with a relatively early English name, bounded by a stream with an apparently British name. Is it possible that, on this low-value land, a small British community was allowed to perpetuate older

land-use patterns here, after the arrival of English-speaking settlers in the dale?

There was nothing new that the Anglian intruders could teach the British peasants about farming; whether the holder of a piece of land in the Dark Ages was a Saxon, a Briton, a Celtified Saxon or an Anglicised Celt, it seems highly likely that the fields which remained in use throughout much of Yorkshire were essentially the ones farmed in Roman times. As Williamson remarks, *à propos* the myth of the Saxon settlers as the makers of the English landscape: 'It is now clear that regular open-field agriculture on the Midland

The ironwork on the church door at Stillingfleet appears to depict a Viking ship with a steering oar (in the lower right portion of the photograph). The door is framed by unusually fine Norman carved masonry – and the Normans did use similar ships to the Vikings

pattern did not exist in Continental Europe in the fifth and sixth centuries: it could, therefore, hardly have been introduced into England, fully formed, by the first Saxon settlers.' One finds the best contemporary examples of working countrysides from post-Roman Britain in Essex, Devon or Wales, where 'ancient countryside', characterised by small, thickly hedged fields, deep, winding lanes, and scattered farmsteads and hamlets, can still be seen. For many parts of Yorkshire, surviving documents and other evidence provide a reasonable understanding of what the medieval countryside looked like, but the appearance and management of the countryside in the centuries between the end of Roman rule and the Norman conquest are infinitely more perplexing. It is plain that the second half of the Dark Ages experienced a substantial, if partial, recovery from the demographic and agricultural catastrophes of the fifth and sixth centuries, and that a remarkable transformation spread across the face of rural England in Viking and later Saxon times, though it is difficult to specify the causes and the nature of this change. What is known is that, firstly, in England as a whole, the recovery from the afflictions of the fifth and sixth centuries was sufficient by the eighth, ninth or tenth centuries to allow the introduction of new systems of open-field farming. Wherever there was ploughland in some quantity and estates that were still large enough to accommodate the new system, then open-field farming appeared. Meanwhile, on small estates and those with only small pockets of ploughland, ancient countryside with the old field patterns endured. Secondly, some of the Yorkshire open-field systems that are described in medieval documents could have been created only by a wholesale reorganisation and planning of the estate in which they formed. Consequently, one might argue, the new rural lay-outs would have been introduced at a time when drastic changes in ownership and rural society created the fluidity necessary for the old traditions and privileges to be overturned. One can recognise two occasions on which the circumstances could have been favourable. The first occurred at the time of the Viking invasion and settlement; the second during the reorganisation of Yorkshire under new masters after the terrifying Harrying of the North by the Norman armies in 1069–70. It is virtually certain that, in Yorkshire as in other parts of England, open-field farming systems appeared in the closing centuries of Saxon rule. The question that is still to be answered concerns whether the planned field lay-outs described in medieval documents appeared – like, as most imagine, the planned villages of the Vale of York appeared – after the Norman harrying, or whether they already existed before the Norman conquest?

In this, and in other respects, the contribution of Scandinavian settlers to the evolution of the Yorkshire landscape is still uncertain. The Vikings were first noted in England in 787, when an official of Beorhtic, the West Saxon king, was murdered when he went to meet a newly arrived band of foreign mariners. Six years later, the Vikings launched their first major assault on English Christianity and culture by destroying the church and community on Lindisfarne. Societies well aware of the domestic capacity for treachery, murder and looting soon learned to reserve a special category of terror and revulsion for the pagan raiders. In the course of the ninth century, the sudden raids on rich but poorly defended monasteries gave way to more prolonged invasions and armies of occupation. There is no doubt that the Scandinavian raiders unsettled affairs throughout eastern England, and that they were greatly feared. The raids on Lindisfarne in 793 and Jarrow in 794 appear to have disrupted the economy, since the minting of coins ceased in the reign of Eardwulf (796–c. 810). Haldenby has described how a hoard of high-quality metalwork that was discovered in a relatively impoverished locality in the Yorkshire Wolds could well represent the flight of English refugees from a more affluent setting which was exposed to Viking raiders. In 865, a large Danish army overwintered in East Anglia;

A human figure, perhaps an apostle, depicted on the shaft of a tenth-century cross at the ancient minster church of Stonegrave

the following year, Ivar the Boneless led a fleet up the Ouse to capture the Northumbrian capital of York (*Eofor-wic*). In 867, Northumbria was subjugated by the great army; the English fought to recover York, failed, and saw the installation of an English puppet king. About the middle of the next decade, York, now *Jorvik*, became the capital of a Danish Viking kingdom ruled by Halfdan, and the *Anglo-Saxon Chronicle* tells us that in 876: 'Halfdan divided the Northumbrian lands and his men were ploughing and tilling them.' This probably means that the Vikings gained control over many estates in the part of Northumbria which would eventually resurface as Yorkshire.

For almost a century, and under a succession of more than a dozen Viking monarchs, York remained a Scandinavian focus. The Danish city was taken by the Norse or Norwegian Vikings in 919 and was Viking-dominated until 954, when Eric Bloodaxe was driven out of York by the English and slain at the battle of Stainmore, allowing Eadred to include this great northern

The turbulence of the Dark Ages is echoed in the lowest panel of the ninth-century Cockshaft cross at Brompton church, where 'beetle-men' are in combat. A more pastoral side to life is reflected by the cockerels in the panels above

England generally, however, the Vikings have left remarkably little in the way of distinctive tangible monuments, and one is forced to wonder how great the Viking heritage in Yorkshire really is. Plainly, it had its destructive and constructive aspects, for, while contributing to northern commerce, urbanisation, culture and dialect, the Scandinavians were also responsible for destroying the legacy of learning, organisation and buildings painfully established by the Benedictine monks. In terms of numbers of people, the Viking contribution to the population of Yorkshire was probably fairly modest. Once again, one has the impression of a smallish élite dominating a large indigenous population before being absorbed into its ranks. The entry quoted from the *Anglo-Saxon Chronicle* for 876 could be translated as: 'this year Halfdan shared out the lands of Northumbria and they ploughed it and made a living for themselves', and it probably gives an

Before the Danish invasions, Hackness supported a nunnery or double monastery. The Hackness cross may be a memorial to the early eighth-century abbess, Oedilburga. The monastery perished around the time of the plundering of Whitby Abbey, 867–9. The cross consists of the upper and lower portions of a taller monument

prize in his kingdom for the final year of his reign. By this time, however, Viking power had waned and the *Jorvik* kingdom was little more than a principality of Eadred's realm. This was not the end of Viking influence in Yorkshire. Harold, the last English king, placed Yorkshire under the control of his brother Tostig, who was soon to be banished for crimes against his subjects. The royal brothers were partly Danish, and Tostig fled to the court of the Norwegian king, Harold Hardrada. In 1066, the fleet of Hardrada wasted the Yorkshire coast and sailed up the Ouse to Riccall, disembarking an army which marched on York. This great Viking invasion force was defeated by Harold's army at Stamford Bridge, after which the English marched south to meet the Norman forces at Hastings.

Much has been written about the Viking legacy in Yorkshire, and the opening of the magnificent Jorvik Viking Centre in 1984, displaying conserved sections of tenth-century timber buildings and a full-size reconstruction of their original appearance, stimulated popular interest enormously. In exaggerated impression of the strength of the Viking presence in the countryside. Place-names, which must always be treated with great caution, suggest a substantial but uneven Scandinavian presence: working with place-names which appear in Domesday Book, one finds that the Scandinavians contributed 40 per cent of names in the old East Riding, 38 per cent in the old North Riding and only 13 per cent of those in the old West Riding. Even so, these names show only that the Viking name for a place has survived, not that the Vikings *founded* the settlement: Scandinavians could easily rename an existing settlement. Also, while many Scandinavian words were absorbed into the Yorkshire dialect, the speaking of Old Norse and Old Danish here did not endure for very long. There is a school of thought which holds that English settlements taken over by Vikings could combine a Scandinavian first element with the English -tun ('a vill') ending. Malton is a Scandinavian name, the 'middle vill or farm', while Flockton, near Dewsbury, is the Viking Floki's vill; Nawton belonged to Nagli, and Sproxton to Sprot. Later Scandinavian settlements could have names ending in '-by', a farm or village, or '-thorpe', an outlying farm. The place-name evidence is anything but straightforward, however, and it is known that some apparently Scandinavian place-names were allocated well after the Viking period was over, using Scandinavian words which had entered the Yorkshire dialect. For example, Godwinscales, near Ripley, is the *skali* or 'outlying hut' of one Godwin, who is known not to have lived until the mid-twelfth century.

Members of the Viking host who settled in Yorkshire after the victory of 867 would almost invariably have taken native wives, though there may also have been immigration by Scandinavian families and settlement by traders of other nationalities during the heyday of *Jorvik*. The presence of Viking women in York is suggested by the discovery, during the Coppergate excavations, of a sock made by the 'needle weaving' technique, a Scandinavian craft which failed to take root in England, while the Fryston place-name between Selby and Castleford could indicate a community of Frisian settlers. (It is interesting to think that, since there was an overlap between the areas on

the North Sea margins which had contributed Saxon settlers in the fifth and sixth centuries and Viking raiders in the ninth, there must have been some places in Yorkshire where Viking and Saxon people shared the same homeland.) The discovery of Anglian York has shown that the Danes were not responsible for the revival of urban life in York and that the town had probably existed as the Northumbrian capital since about the time of Edwin's baptism there in 627. Nevertheless, the incorporation of York into the Viking trading arena must have led to growth and to an enormous quickening of its economic life. For a few decades, the Viking-held territories were virtually removed from their earlier contexts and integrated into the vigorous world of Scandinavian commerce. York and Dublin were two great trading centres in Viking Britain, linked by water routes, via the Ouse and the North Sea, and by portage, with a highly important portage route across the Forth–Clyde isthmus in Scotland connecting the North Sea and Irish Sea shipping lanes. Trans-Pennine routes, notably via Wharfedale and Ribblesdale and the Roman York–Carlisle road, linked Danish York to the Norse Viking territories in Galloway, Cumbria and Lancashire and, thence, via the Irish Sea, to the Viking trading towns and military bases in Ireland.

More than at any time since the Roman collapse, York enjoyed a leading role in an important international commercial arena. To this day, many of the city's streets preserve the Danish ending '-gate', meaning 'street': Petergate, Davygate, Fossgate, Monkgate, Micklegate, Stonegate, Gillygate, Walmgate, and so on, Skeldergate, for example, being the street of the shield-makers. The main commercial district of Viking York lay to the south-east of the Roman nucleus, in the angle between the Rivers Ouse and Foss, guarded by water and lined with wharves. Within this area, the Coppergate excavations explored a street populated by artisans who shaped wooden cups and bowls ('Cupper-gate' denotes the street of the cup-makers), cured leather, and shaped beads from amber. Longships at the waterfront bore oriental silks, Siberian furs from the trading sphere of the Swedish Vikings (Varangians), wine from the Baltic, slaves from Ireland and plunder from many places. Yet, for all its bustle and vibrancy, *Jorvik* was a world away from the elegance and fastidiousness of Roman *Eboracum*: during the Dark Ages, some nine metres of garbage and filth accumulated above the Roman surface. Had the Vikings shown a higher regard for cleanliness and order, however, our archaeological knowledge of them would have been much less (while, had the decay and destitution of Roman York been less, the Viking streets would have shown a closer correspondence to the Roman alignments). One cannot deny the momentum

---

The resurgence of Christianity in Yorkshire is symbolised by the rebuilding of decayed minsters, like St Gregory's, at Kirkdale. This may mark the site of the seventh-century monastery of St Cedd, though Lastingham is more frequently regarded as the site of 'Lestingau' monastery, where the Celtic custom of Lindisfarne was followed. At the synod of Whitby, St Cedd agreed to adopt the Roman custom. The seventh-century church was destroyed by Danish raiders and lay in ruins until it was rebuilt by Orm, son of Gamal, in 1055–65

that Viking control gave to the renaissance of York, even if the city's population was probably composed of indigenous people under Viking leadership. In terms of tangible Viking relics, however, the heritage consists largely of monuments that are Scandinavian in style yet which appeared within the context of the established Christian Church. In 878, Alfred defeated the Danes at Edington and Guthrum accepted Christian baptism. Gradually, for reasons of expediency or sincere conversion, the Vikings in England renounced paganism and accepted the Christian faith, so that Scandinavian artistic motifs and mythology came to be preserved in churches, tombs and crosses. One type of monument that is usually associated with Scandinavians is the hogback tombstone, a massive grave-slab arched like the back of a boar, which is thought to have the generalised form of a contemporary house. One of England's finest collections of these monuments was kept in the church at Brompton, near Northallerton. Some have been moved to the library of Durham Cathedral, but three excellent examples remain, with the tombstones held

Kirk Hammerton church also exemplifies the revival of Christian worship after the pagan Vikings. Some experts regard this as a very early church, with a 'timeless' and deceptive appearance, while others see it as a monument from the end of the Saxon church-building tradition

in the clasp of stone bears. Other examples can be seen in the church at Burnsall, in Wharfedale, at Pickhill, near Thirsk, and at All Saints, Dewsbury. Scandinavian artistic influences can also be recognised in several carvings: the intertwined men and beasts and interlace on the cross fragment at West Marton church, near Skipton; the animal ornament on the wheel-headed cross in the church of North Frodingham, near Great Driffield; the carved wheel-headed crosses at Middleton church, near Pickering; and similar wheel-headed crosses in Lower Swaledale, at Stanwick, Gilling and Finghall. The Scandinavian rural settlements remain largely mysterious, perhaps because they cannot be distinguished from those of non-Scandinavians resident in Yorkshire. The best candidate must be the Ribblehead site, near Ingleborough, where the ruins of a long, stone farmhouse have been dated to around 870.

The inhabitants were farmers who used limestone slabs to build their stockpens and who also engaged in metalworking.

The Vikings made a contribution to Yorkshire dialect and might have provided the broad vowels associated with the accent. They provided many place-names, although the folly of confusing names with new settlements has been noted: if every village with a Viking name had been home to a village-sized Scandinavian community, then Yorkshire would have been awash with Vikings – and this was surely not the case. Take, for example, the village of Ramsgill in Upper Nidderdale. The name includes the Old Norse '-gil' ending and probably means 'the gill or ravine of the wild garlic'; the 'Ram' element probably derives from the Saxon *hramsa*, meaning 'wild garlic' or 'ramsons'. Close to Ramsgill is Bouthwaite, with a Scandinavian name which probably means 'cottage in the meadow or clearing'. There is no reason to suppose that either of these places were Viking villages. They first appear as granges or outlying farms of Cistercian abbeys which eventually grew into hamlets; only in the post-medieval period did Ramsgill emerge as a small village. Many of the Scandinavian place-names which now attach to villages were originally descriptions not of settlement, but of local topography: *gil*, *-ey*, an island; *kjarr*, a bushy bog; *myrr*, a marsh; *steinn*, a rock; *fjall*, a fell; and so on. Even the names that do refer to settlements usually denote farmsteads rather than villages: '-by', '-toft', and words like 'saetr' and the Norse-Celtic 'erg', which denote shielings. Resuming the Nidderdale example, the dale is packed with settlements with Viking names: Lofthouse, Ramsgill, Coldstones, Fellbeck, Thruscross, Thornthwaite, Hartwith, Tang, Graystone, Clint, Bouthwaite, and many others – but not one of them necessarily denotes a village or hamlet *founded* by Scandinavian settlers. All that such names reveal is that the Old Norse and Old Danish languages made a heavy contribution to the English dialect of Yorkshire and that Scandinavian words were often used to describe places, some of which acquired villages at one time or another. Faull writes that: 'Until there is more certainty about whether the early place-names in the first instance refer to specific points in the landscape or to units, they cannot be used for identifying pre-Conquest settlement sites. This is certainly the case with names recorded in Domesday Book, which relate to vill units not individual settlements.'

The same sorts of problems are encountered where English names are concerned, and major revisions have utterly transformed the very foundations of study. It was held that oft-encountered names incorporating the '-ing' element, like Dunnington, Heslington, Riplingham, and so on, were used early in the Saxon period and denote the arrival of Saxon leaders and their *ingas*, people or followers, so that Fylingdales would be 'the valleys of Fygla's people'. Equally, it was said that '-ley' place-names, like Bramley, Keighley, and so on, came later, because they showed how communities from the original Saxon settlements were now becoming daughter settlements in new *leah* or clearing places (a name like Headingley

presents problems for a schema such as this!). In fact, the common '-ley' names demonstrate the difficulties of place-name evidence, for while 'leah' can denote a clearing, it can also mean a 'wood', or an area of open pasture. Names which end in '-ton', denoting a Saxon *tun*, a farmstead or vill, were also presumed to be secondary, like the '-ley' names. In the eastern areas of early Anglian settlement and pagan burial, however, are found countrysides littered with supposedly secondary '-ton' names: Seaton, Brandesburton, Boynton, Bainton, Etton, and many more. The current wisdom argues that Saxon '-ham' place-names are very early, '-ham' coming from *ham*, meaning a 'homestead' or 'settlement', and holds that '-ing-ham' and '-ing' names are next in antiquity. (If this is true, then Kirkham, 'church homestead', near Malton, would seem to be a problem, for the 'Kirk' element ensures a post-conversion date. Also, the '-ham' names do not correspond very well to the pattern of pagan Saxon cemeteries in

The nave and mutilated and repaired Saxon arch at Kirk Hammerton

Yorkshire.) According to Gelling, the '-ton' and '-ley' names belong to a middle phase of English settlement, roughly to the period 750–950. The correct interpretation of a name is often impossible; the more reliable translations combine expertise and local knowledge with an effort to choose a translation that accords with the history and topography of the place concerned. The opportunities to make mistakes are legion. To note just one example: overlooking Nidderdale there is a locality known as 'Hark Hill' and a farm called 'Hark Hill Nook'. The records show that, in the 1340s, the neighbourhood was cleared of woodland by one Henry Arkel, so it would be easy to imagine that this coloniser gave his name to the land. However, in the reign of Edward the Confessor, the locality concerned was part of the estates of an important nobleman called Archil. Presumably, it is Archil's name which has been preserved in the countryside, and Henry Arkel was simply Henry of Arkel.

To the landscape historian, place-names can be useful when they describe how a location once looked, with names like Swinden describing 'the valley of the pigs', Rathmell recording 'the red sand bank', Doncaster 'the old Roman station on the Don', and Harrogate perhaps 'the road or right of pasture by the heap of stones'. The names also serve as reminders of how cosmopolitan a place Yorkshire once was, and how mixed its population. It is conceivable that, at the time that Elmet was overrun by the Anglians, there could have been a few educated men there who could converse in at least one Celtic language, Latin and Old English; while around the tenth century, peasants living in the far north-west of the region might have understood the Celtic tongue still current in Cumbria, Old Norse, Old Danish and Old English. A few place-names derive from the Old English *wealh*, denoting foreigners or 'Welsh' people, and these may indicate communities of British people – Walden, off Wensleydale, could be an example, signifying 'the valley of the British'. There are other names which imply a mixing of the ethno-cultural strands. Yockenthwaite in Wharfedale suggests an overlap of Norse and Celtic speakers, for it means the 'thwaite', the Norse word for 'a clearing' or 'meadow', belonging to the Celt Eogan. Pockets of Scandinavian-speaking people might still have been present in Yorkshire at the time of the Norman conquest. The dialect and countryside of the dales is particularly rich in words of different origins, and it was here that Norse immigrants crossing the Pennines from the west mingled with Danish speakers moving in from the east. So, for example, one finds in Nidderdale that the River Nidd has kept its Celtic name ('brilliant water'); and that people from Dacre (the Celtic 'trickling stream') look across the Nidd to Hartwith (the Old Norse 'wood of the stag'), or turn southwards to Darley (the Old English 'glade of the deer'), or beyond to the Danish-settled area, which they meet in parts of the Vale of York to the east. Even so, the names provide an endless source of debate; Dacre, for example, might just come from the medieval Latin *Dacus*, 'a Dane'. In the centuries preceding the Norman conquest, Yorkshire must have existed as a cultural melting-pot, and students of linguistics have shown how English was 'pidginised' to facilitate understanding among people of varied ethnic origins. McCrum, Cran and MacNeil explain that:

> Before the arrival of the Danes, Old English, like most European languages of the time, was a strongly inflected language. Common words like 'king' or 'stone' relied on word-endings to convey a meaning for which we now use prepositions like 'to', 'with' and 'from'. In Old English, 'the king' is *se cyning*, 'to the king' is *thaem cyninge*. In Old English, they said one *stan* (stone), two *stanas* (stones). The simplification of English by the Danes gradually helped to eliminate these word-endings . . .

They also note how rapidly the assimilation of the Vikings took place; writing of Kirkdale church, they explain that: 'Lovingly chiselled into the stone is the resoundingly Viking name of the man who made it: Orm Gamalsson. But on closer inspection the inscription turns out to be worked in Old English, not Old Norse. Barely one hundred years after his people had invaded Britain, Orm Gamalsson is writing (and presumably thinking) in English.'

There is one more facet to the Viking heritage: the Vikings were at least partly responsible for providing the county of Yorkshire, for it seems that the three ridings (i.e. 'thirds'), the West, North and East, were created in Viking times, with the East Riding holding its Parliament or 'Thing' at *Jorvik*. Moving down to the next administrative level, the Scandinavian division into wapentakes gained ascendancy over the Saxon hundreds, which were normal in most other parts of England. Yorkshire itself derived from the Viking kingdom ruled from York and was absorbed into the English kingdom intact after the vanquishing of Eric Bloodaxe in 954. The word 'shire' comes from a Saxon word for a division of land, and the name 'Yorkshire' is first recorded in 1055, although it would have been in use for a while by then.

History had been hinting at the existence of Yorkshire polity for a long time. Brigantia had been too large, the territory of the Parisi too small; Deira had also been rather too small – and Northumbria rather too large. Like the Romans, King Edwin had recognised the geographical and strategic importance of York; it became his Northumbrian capital and, later, the capital of the Northumbrian territory captured by Halfdan's Danes. Now, in the closing decades of the Saxon period, Yorkshire and its thirds had gained formal recognition. It was a place of mixed cultural origins, markedly less 'English' than most of the lands to the south.

REFERENCES

R. N. Bailey, *Viking Age Sculpture in Northern England* (London: 1980).

S. Bassett, 'In search of the origins of Anglo-Saxon kingdoms', in S. Bassett (ed.), *The Origins of Anglo Saxon Kingdoms* (Leicester: 1989), pp. 3–27.

M. L. Faull, 'Late Anglo-Saxon settlement patterns in Yorkshire', in M. L. Faull (ed.), *Studies in Late Anglo-Saxon Settlement* (Oxford: 1984) pp. 129–42.

M. L. Faull and S. A. Moorhouse (eds), *West Yorkshire: An Archaeological Survey to AD 1500*, 4 vols (Wakefield: 1981).

A. Fleming, 'Swadal, Swar (and Erechwydd?): early medieval polities in Upper Swaledale', *Landscape History*, 16 (1994), pp. 17–30.

M. Gelling, *Place-Names in the Landscape* (London: 1984).

D. M. Hadley, 'Multiple estates and the origins of the manorial structure of northern Danelaw', *Journal of Historical Geography*, 23 (1996), pp. 3–15.

D. Haldenby, 'An Anglian site on the Yorkshire Wolds', *Yorkshire Archaeological Journal*, 62 (1990), pp. 51–63.

R. A. Hall (ed.), *Viking Age York and the North*, Council for British Archaeology Report 27 (London: 1978).

D. Hey, *The Making of South Yorkshire* (Ashbourne: 1979).

N. Higham, *The Kingdom of Northumbria* (Stroud: 1993).

J. Hines, 'The Scandinavian character of Anglian England in the pre-Viking period', British Archaeological Report, British Series No 124 (Oxford: 1884).

G. R. J. Jones, 'Early territorial organisation in northern England and its bearing on the Scandinavian settlement,' in A. Small (ed.), *The Fourth Viking Congress* (Edinburgh: 1965), pp. 67–84.

G. R. J. Jones, 'Multiple estates and early settlement', in P. H. Sawyer (ed.), *English Medieval Settlement* (London: 1979), pp. 9–34.

R. McCrum, W. Cran and R. MacNeil, *The Story of English* (London: 1987).

D. J. H. Michelmore, 'The reconstruction of the early tenurial and territorial divisions of the landscape of northern England', *Landscape History*, 1 (1979), pp. 1–9.

J. Thirsk, 'The origins of the common fields', *Past and Present* 33 (1966), pp. 142–7.

R. L. Thomson, 'Celtic place-names in Yorkshire', *Transactions of the Yorkshire Dialect Society*, 2/64 (1964), pp. 41–55.

T. Williamson, 'Explaining regional landscapes: woodland and champion in southern and eastern England,' *Landscape History*, 10 (1988), pp.5– 13.

CHAPTER NINE

Domesday Countrysides of Yorkshire

THE DOMESDAY SURVEY WAS COMPILED in 1086 and it catalogued a land which was still partly devastated by the ravages of the Norman host in 1069–71. In 1066, the Norse army of Harold Hardrada and Tostig had left its boats at Riccall and marched on York; at Fulford, it met and defeated an English army led by the northern earls, Edwin and Morcar. The invaders were attacked and defeated by King Harold's forces at Stamford Bridge, but within three weeks Harold, with the best of his southern warriors, lay dead on the battlefield of Hastings. The news of this defeat would soon have found its way to Yorkshire, where its true implications are unlikely to have

been appreciated at first. The mourning for Harold, able king though he had been, may have been muted: northerners included many people of Scandinavian descent; they could look back to the days of an independent Northumbria, and they would not welcome strong rule by monarchs based in the south. Although William was crowned at Westminster Abbey by Aldred, the Archbishop of York, and received homage from some of the

The Baile Hill motte at York, perhaps built by William the Conquerer during his first attack on York and later traversed by the city wall

Clifford's Tower at York stands on an earlier Norman motte

Yorkshire earls, the finality of the defeat for the Anglo-Danish monarchy was not accepted in the North. In 1068, Edwin and Morcar renounced their allegiance to the Norman king; William marched on York, hurriedly erected some fortifications, perhaps including the mound or 'motte'-known as Baile Hill, and installed a garrison under the sheriff, William Malet. At the start of 1069, Robert de Comines was created Earl of Northumbria after Copsige, the previous appointee, had been killed when he tried to enter his earldom. Robert arrived in Durham at the head of a substantial army – all of whom were killed in a new revolt, after which the Northumbrians marched southwards. The revolt then flared again in Yorkshire; Edgar Atheling, of the lineage of Alfred, was proclaimed king at York, where the Norman commander was killed and his men besieged in their castle. William returned to raise the siege, erected another castle mound – probably the one which now carries Clifford's Tower – and massacred large numbers of the city's inhabitants.

In the autumn of 1069, Sweyn of Denmark raided Kent and East Anglia and then took his fleet up the Humber. This time the Scandinavians were welcomed as allies and York became a rallying point for dissidents from many parts of England, southern Scotland and Wales. Faced with this formidable confederacy, the frightened Norman garrison burned York and its old minster, but, with victory in their grasp, the allied armies fragmented, some Danes returning to their ships with their loot, leaving the Northumbrians to fend for themselves. William rode from his campaigns in the west of England at the head of a large cavalry force, determined finally to crush Northumbrian resistance. The English-dominated lands in Lothian were commended to William's vassal, Malcolm of Scotland, and the remaining Danes were bribed to return to their homelands. William then ordered the systematic campaign of genocide and destruction known as the Harrying of the North. The near-contemporary historian Ordericus Vitalis recorded that some 100,000 northerners perished in Yorkshire, Lancashire and the southern parts of County Durham, and he described how those who survived the killings then faced famine in their ravaged countrysides: 'corn, cattle and every kind of food he ordered to be heaped together and burnt. The famine that had already raged for more than a year was aggravated by such means and so horrible was the misery that the wretched survivors were forced to exist on horse meat, cats and even human flesh.' He continued to describe a land wreathed in the stench of putrefaction and strewn with unburied corpses: 'Nothing moved amongst the ruins of the burnt-out villages but packs of wolves and wild dogs tearing at the bodies of the dead.' Then Malcolm's armies appear to have invaded from the North, killing indiscriminately, looting, burning and returning to Scotland with masses of English slaves. According to the chronicler Simeon of Durham, not a single inhabited village remained between York and Durham, and the land there lay uncultivated for nine years.

This was a period during which violent death was almost commonplace and martial skills, intransigence and ruthlessness towards enemies were qualities to be valued in rulers. Even so, there is good reason to believe that contemporaries who were accustomed to campaigning and violent retribution were shocked by the atrocities of the Harrying. As a boy, William the Conqueror had survived in conditions of intrigue and almost constant danger – he had already learned the effectiveness of scorched-earth policies during his continental campaigns against Maine – but the scale and thoroughness of his massacres in Yorkshire shocked even his natural sympathisers,

like Ordericus Vitalis: 'I have been free to extol William according to his merits, but I dare not commend him for an act which levelled both the bad and the good, in one common ruin by a consuming famine . . . I assert moreover that such barbarous homicide should not pass unpunished.' He claimed that, on his deathbed, William had repented 'the death of thousands by starvation and war, especially in Yorkshire'.

The study of history sometimes proceeds in a cycle of debunking and rehabilitation, so that any event or figure that seems 'larger than life' will experience a severely critical re-evaluation. This has been the case with William's Harrying of the North, but, fortunately, an objective assessment has recently been provided by Palliser, who set out to evaluate Freeman's claim of 1871 that 'William had made a wilderness and he called it peace'. Considering the accounts of the medieval

The Norman motte-and-bailey castle at Barwick, recently cleared of encroaching scrub. The design was unusual, with the motte placed at the centre of the bailey, perhaps because the Norman master commandeered a corner of a much larger Iron Age hill-fort. Political authority may thus have been centred at Barwick for many centuries

chroniclers, Palliser noted that Ordericus was born in 1075 near Evesham and could have been writing in 1125, some 55 years after the events described, and that his figure of 100,000 victims was used in a rhetorical sense – he also attributed the massively inflated figure of 60,000 knights to William. He was writing in Normandy and knew little of the geography of the northern counties concerned. William of Malmesbury, who claimed that, because of the Harrying, land was left untilled over an area of more than 60 miles and that the area still lay uncultivated, was also writing 55 years after the events. Simeon of Durham simply amplified an account by Florence of Worcester that was written in the early twelfth century. However, although there are uncertainties about the degree to which the devastation was caused by Scottish raiding rather than Norman harrying, there is independent archaeological evidence that violent disruption took place in Yorkshire in 1069–71, in the form of hoards of coins which were buried by the inhabitants. Palliser judged that: 'It would be wrong to conclude that William's

The earthworks of the early Norman fortress of Skipsea, near Hornsea. The great motte (upper right) was separated from the encircling ramparts of the bailey by a tidal mere and was reached via a wooden causeway

punitive actions in 1069–70 were not catastrophic . . . it was clearly a harrying which shocked men of the twelfth century and struck them as beyond normal or acceptable limits.'

William spent the Christmas of 1085 at Gloucester, where the idea of the Domesday survey was conceived. As Palliser points out, the significant facts about the Domesday record for Yorkshire are its omissions and its unreliability. The total population recorded for Yorkshire in 1086 is only 7,570, which can be interpreted as a population of only three per square mile – one of the thinnest in England and a figure which must massively underestimate the actual number. According to Holt, the Yorkshire entries in Domesday seem hurried and artificial, suggesting that the king may have applied pressures for urgency. This could have caused particular problems for compilers working in distant northern counties: some may have had only six months to collect and record the data before a deadline of 1 August 1086. When the commissioners arrived in Yorkshire, it is clear that they found a land that was still crippled by the campaigns of 1069–71. The entry for estate after estate reads 'wasteas est' or 'hoc est vasta': it is wasted. According to Maxwell, Domesday Book shows that, even 15 years after the Harrying, 480 of Yorkshire's 1,782 vills were still waste and 314 more were partly waste; in the North Riding, two-thirds of the

estates were still either waste or partly waste. There are, however, considerable problems of interpretation here. 'Waste' was a term that was sometimes applied to commons, and it is not clear whether it was used in Domesday to describe formerly productive land that had gone out of cultivation. Wightman believed that 'waste' in Domesday applied not only to abandoned or worthless land, but was also used to denote manors that had been amalgamated or were in disputed ownership. However, Harrison and Roberts, two leading authorities in northern landscape history, are quite specific; with regard to the Pickering estate they write: 'One thing is, however, chillingly clear: in 1086 the estate was producing only a fraction of the revenue it had generated in 1066, and the phrase "The rest (is) waste" provides a clear explanation: *waste* was land which had been devastated.' When one maps the pattern of wasted vills and manors, it seems that only the loftiest parts of the Pennines, the plateaux of the North York Moors, and much of Holderness escaped severe devastation, while the worst of the wasting seems to have taken place in the Pennine foothills and the Dales. In Nidderdale, for example, 13 of the 14 estates were waste. But the pattern of 'wasted territory' does not seem to correspond to the route which a force of invaders might be expected to have taken. Their effectiveness would surely have been greatest in the lower, flatter countrysides of the Vale of York, which offered ease of access, more targets and fewer hiding places. Bishop's suggestion was that those members of the new aristocracy who had estates both in the Dales and the Vale of York had, between the massacres of 1069–71 and the compilation of Domesday in 1086, moved survivors from their homelands in the Dales and foothill country to repopulate the potentially more affluent estates of the Vale of York, the places which might well actually have borne the brunt of the Harrying. This thesis might be impossible to prove and it has received only limited support. Hey also had problems with the relationship between waste and the Harrying. He wrote:

> Little or no waste is recorded along such major invasion routes as the lower reaches of the Ouse or the approach to Stamford Bridge from

Bridlington, while on the other hand plenty of waste vills are recorded in Holderness and parts of south Yorkshire, which lay well away from the areas that suffered from the harrying . . . Could a comparatively small army really have done sufficient damage for so much land to remain devastated seventeen years later? Can we really believe that no less than 480 settlements, that is 27 per cent of those recorded in the Yorkshire folios of Domesday Book, were deserted for at least seventeen years, yet nearly all of them were subsequently resettled with their former place-names and boundaries intact?

Most of the old nobles were dispossessed and their lands were divided between the king and 26 of his tenants-in-chief (who held their land directly from the king). The Count of Mortain, the king's half-brother, held no less than 215 manors in Yorkshire, and other large landowners included Count Alan of Brittany, Roger de Poitou, William de Percy, William de Warenne, Roger de Busli and Ralph de Mortimer. In many places, apparently, the depopulated and devastated estates would often have been suitable only for hunting. In the more rugged parts of the Dales, land often remained under hunting forest throughout much of the medieval period. In Upper Wensleydale, the forest of Bainbridge spanned two wasted vills; Nidderdale supported the chase of Nidderdale, soon to be assimilated into Cistercian estates, and the more durable royal Forest of Knaresborough; at the head of Swaledale and Wensleydale were the Forests of Swaledale, Wensley and Mallerstang; and the forest of Litton and Langstrath covered the head of Wharfedale. Some of these areas would already have existed as hunting forest, perhaps parts of the great forest which was mentioned in the reign of King Cnut. Hints of such a forest come from Swaledale, where it is known that the lords of Richmond hunted in the Forest of Arkengarthdale and the 'New Forest', suggesting that the latter was an addition to an already existing hunting territory. In the lowlands, however, the more fertile farmlands promised higher rents and profits, so the emphasis shifted from destruction to rehabilitation.

The field systems that developed in Yorkshire

Vast areas of the Yorkshire Dales, like Upper Wensleydale, shown here, existed as thinly populated hunting forests under the Norman kings

in the course of rural recovery were varied and complex. Here, as in many other parts of England, it appears that there was a wholesale reorganisation of the lowland estates during the late Saxon or Viking period. Where the existing estates were still large enough to accommodate the new model, there was a division of land into pasture, meadow, common or waste (which was not, by any means, worthless land and could include valuable woods or heath), and communally worked ploughland, which existed as strips scattered amongst the blocks or furlongs in vast, open (that is to say, unenclosed or partitioned) fields. In southern and central England, there is good evidence for the establishment of open-field farming well before the Norman conquest. Hooke notes that: 'Charters of the tenth and eleventh centuries refer to arable strips in common fields, to intermingled holdings and to related holdings in the manorial waste in such a way as to leave little doubt that many of the features characteristic

of medieval farming in the twelfth, thirteenth and fourteenth centuries had already come into being.' Such open-field systems were probably established in Yorkshire, but it is not clear whether or where they survived through the Harrying; some of the systems described in the documents of the later centuries show clear evidence of planning, which could have taken place when the Norman lords were seeking to rehabilitate their devastated estates. It is important to understand the main land divisions employed in medieval Yorkshire. These were the 'bovate' (or 'oxgang'), a variable unit generally of 8–15 acres of ploughland, plus an acre or two of meadow and common grazing rights, and the 'carucate', notionally the amount of land that an ox-team could plough in a year and equivalent to eight bovates. It was common for a peasant tenant in Yorkshire to have a two-bovate holding, though holdings of multiples or fractions of bovates could also exist. The bovates were often fragmented amongst the village fields in 'selions' or 'lands', which we now refer to as strips, though sometimes the bovate holdings formed more compact blocks.

When Domesday was compiled, the commissioners sought to answer the following questions for each estate. For how many ploughlands or carucates is the place taxed, and how many ploughs are there? Who has the estate now? How many peasant tenants of the 'villein' and 'bordar' grades does he have? How much pasture or woodland is there? What was it worth in King Edward's time, and what is it worth now? The answers were recorded in a form of Latin short-hand, and it is worth emphasising that they do not necessarily tell us anything about villages and that other contents, like priests and churches – or, occasionally, even a castle – could be over-looked. In Yorkshire, as noted above, the gather-ing and recording of Domesday data appear to have been particularly wayward. Some of the entries are very short and uninformative, but the three examples which follow are longer and show how the formula was applied:

Manor. In Studley [near Ripon] Ledwin has 13 oxgangs [or bovates] of land to be taxed, and there may be one plough there. Archil now has it of William, and it is waste. Value in King Edward's time 10 shillings. *Manors.* In Kirk Hammerton, Turchil, Gamel and Heltor had 6½ carucates of land to be taxed. There is land to 6 ploughs. John, a vassal of Osbern's, has there 2 ploughs, and 5 villeins with 1 plough. There is a priest and a church, and 1 mill of 2 shillings, and 1 fishery of 3 shillings. The whole ½ mile long and ½ mile broad. Value in King Edward's time 4 pounds, now worth 45 shillings. In Easingwold there are 12 carucates of land to be taxed, which 7 ploughs may till. Morcar held these for 1 manor in the time of King Edward. It is now in the King's hands, and there are 10 villeins having 4 ploughs; a church with a priest; wood pasture 2 miles in length and 2 miles in breadth; in the whole 3 miles in length and 2 in breadth. Then it was valued at 32 pounds, now at 20 shillings.

Clearly, the Domesday survey of Yorkshire reveals something of the nature of the country-side and provides highly controversial insights into the more lasting effects of the Harrying, but little is revealed about the form of the fields and the manner of their cultivation. This information must be gleaned from later documents and maps,

and the picture that seems to emerge is one of great variation from one place to another. Harrison has studied the field evidence from charters dating mainly from 1180–1280; field systems were detected in the northern section of the Vale of York and on the Magnesian limestone belt, but, moving south through the Vale of York smaller blocks of ploughland, 'assarts' (areas cleared from the woodland), and woodland were more com-mon. The twelfth-century records told of two 'sides' of the vill, which might indicate that peas-ant communities ploughed two great open fields, with later assarting creating a third field to produce a more conventional three-field system. Within the sides, the individual holdings did not seem to have consisted of the conventional scattered strips, but the sides seem to have been divided into blocks of land or 'cultures' of a carucate or half a carucate in size.

A rather different pattern of farming emerges from Harvey's study of the township of Preston, in Holderness. Here the ploughland was divided between two great open fields, North Field and South Field. In 1750 the land there was recorded as belonging to one William Constable, and it is quite possible that the early medieval pattern was still preserved. Instead of the expected pattern of fields, divided into haphazardly orientated furlong blocks and then into conventional strips, each field was neatly divided into seven divisions called 'bidles' or 'bydales', which were much larger than conventional furlongs, and each bydale was numbered one to seven. The bydales were divided into lands or strips, but these could be very much longer than conventional strips, the longest being a remarkable 2,286 metres. Moreover, the lands within each bydale were referred to by particular names, occurring in a consistent order from one bydale to another; although these names did not refer to the post-medieval landowners, they could sometimes be related to the surnames of medieval occupants of Preston. A careful study of the evi-dence led to the suggestion that, late in the Dark Ages or at some early stage in the Middle Ages, quite possibly after the Harrying of the North, the holdings in the village were laid out in a regular and carefully planned way.

The system of planning employed would be that known as 'solskifte' or sun division, an arrangement discovered in Scandinavia and more recently recognised in a few places in Yorkshire. It ensured that the strips making up a landholding or bovate were arranged in a precise order in every furlong (or bydale) of the open field. The bovate would be divided into regularly scattered strips, so that, in any furlong, the owner of a bovate might have, say, the second most south-easterly strip, which might be described as 'nearest to the sun but one', and the strip-holder would have the same neighbours. In Scandinavia and, perhaps, in Yorkshire, the allocation of the strips corresponded to the position of a peasant's house-plot in the village, so that the occupier of the most westerly plot in the village would tenant the most westerly strip in each furlong. This is obviously a very complicated method of apportioning land, at least when first encountered. But it is also clear that such a system could only be

the result of a complete reorganisation of the land and detailed planning and apportionment. Harrison and Roberts described the operation of a solskifte system in the North York Moors:

the order of strips in a furlong could follow the order of the houses in the village. A good example is at Morton, now Morton Grange near East Harsley. Here Rievaulx Abbey was granted in about 1173 'half of the vill of Morton . . . the carucate to consist of eight perches (one perch equals about 20ft) lying together towards the sun and the tofts and crofts likewise'. This description exactly places the strips making up the 49 ha (120 acres) in a carucate.

The date of this reorganisation of countrysides in Yorkshire to produce planned or 'regulated' field systems is much debated. One school of

thought maintains that it was instigated by the Scandinavian aristocracy and that large areas of the county were fully cultivated before the Norman conquest. Thus it can be argued that, while the Harrying destroyed many people and villages, it did not extinguish the existing field patterns and that these fields were reactivated as the rural communities gradually recovered after the devastation. Another argument asserts that the countrysides were not fully worked until well after the conquest and that the Harrying provided the opportunity for a remodelling of fields and village life, with many new additions to the landscape appearing gradually in the medieval period. A generally accceptable answer would be hard to find and, as Harrison and Spratt point out: 'It is possible that in their time, villages were regulated and re-regulated repeatedly, depending perhaps upon their economic fortunes. To add further

complexity, each village differed from this simple model in response to local circumstances.' Having shown that the Preston pattern of precise planning and a simplified arrangement of ploughland into large furlongs, very long strips and regular allocations between the peasant tenants was characteristic of Holderness in general, Harvey then turned to the Yorkshire Wolds, where she found similar patterns of medieval farming. Other experts working at Wharram Percy in the Wolds were able to recognise the previous existence of a countryside covered in 'long strips', almost all of them aligned from north to south, which were often over 914 metres long and which rolled across the landscape taking little account of the natural terrain.

Harvey was also able to discover some evidence of the introduction of a sun-division pattern in the Vale of York, notably at Fangfoss, where a survey of 1363 showed that the Dean of York always had the strips at one end of a furlong, and at Wressle, where 25 strips separated each glebe strip. While one can appreciate the appeal of the sun-division system of allocation to the organised or bureaucratic mind, the attraction of the long strip is much harder to recognise. More conventional English field strips were around 201 metres in length and 1⅓ acres in area, dimensions which supposedly related to the distance that an ox-team could pull the plough without resting and the amount of land that they could plough in a session. The long strips that have been recognised in Yorkshire and in a few other places in England might be a mile or more in length. We have seen that these long strips were grouped in long furlongs or bydales, and it may be that a very old pattern survived in the east of Yorkshire and that, in most other parts of England, the original long furlongs were broken up to give the familiar, smaller and less elongated furlongs.

An ancient pattern of long strips survived at Middleton, near Pickering, as a consequence of their enclosure by hedgerows. Many of the hedges have now been removed, but the two seen in this photograph trace the outlines of a package of numerous long strips. Note the characteristic slightly curving form and the extreme length of the strips, which continued to, or beyond, the distant trees

Matzat has shown that there are two types of field layout in East Yorkshire:

The first type . . . occurs in the Vale of York and on parts of the Wolds and is of a complex nature: there are a large number of long and short furlongs, running in different directions. The second type . . . which one finds in Holderness and some parts of the Wolds is characterized by extremely long strips (e.g. 10 m wide and 1250 m long), which cover most of the township. This rather simple field lay-out, which for convenience I shall call the 'Holderness type', existed in combination with different village plans.

Initially, Harvey thought that the Holderness type had resulted from a reorganisation of the countryside following the Norman conquest, though in 1983 she revised her thinking in favour of a late ninth-century origin. Meanwhile, Hall, an expert in field-walking techniques, had detected the former existence of similar fields in the East Midlands and favoured an eighth- or early ninth-century origin, when a wholesale rearrangement of rural settlement patterns appears to have taken place. Hall believed that, subsequently, sometime before the thirteenth century, the long strip layouts were reorganised and divided to produce the classic patchwork furlong patterns. Similar patterns to the Holderness type also existed in Germany, particularly in the eastern areas colonised in the eleventh and twelfth centuries. In England, Hall suspected that the introduction of long strips in the middle Saxon period was related to the use of the fixed mould-board plough; this received support from research by Finke at a hamlet site in Westphalia, where such a plough was introduced in the eighth or ninth century and produced a reorganisation of the farming patterns. It had the great advantage over more primitive ploughs of being able to manure the land by ploughing *plaggen*, the uppermost level of heath and other soils, down into the sandy plough-soil. Matzat summarises the evidence as follows:

The research on the genesis of the settlement pattern in the area of Saxon settlement in Northwest Germany exhibits striking parallels for the 8th and 9th century: small settlements were rearranged into larger ones and the older theory, that strip fields originated in the 3rd to 5th century, can be disregarded for both countries . . . The English and German research shows that the origin of long strips may have two independent roots: in one case they are an adaptation to the fixed mouldboard plough, in the other case they are deliberately planned and the result of a simple measurement technique no matter what agrarian impliment was used.

In many upland areas, notably the Dales, ploughland was too scarce to allow the creation of a two- or three-field system with extensive tracts of permanent communal ploughland, and peasant farmers were few in number. In such places, a small community might have operated a one-field system, with the single-strip field, the 'infield', being heavily manured and kept in continuous production, while sections of the surrounding 'outfield' would occasionally be ploughed and the land then rested for several years under pasture. Variations of this system also existed, with one from Snainton in the North York Moors being described by Harrison and Roberts: 'In the thirteenth century a suite of three open arable fields lay on the ridges between the valleys . . . These formed an apparently regular three-field system but there were other open fields nearer the village about which we know very little . . . the fields near the village were regularly-cropped "Infields" and those to the north "Outfields" which were sown less frequently.' Here, too, there was evidence of regulation:

Some of the grants show that these 'Outfields' bear all the marks of a planned layout. Within the individual furlongs, holdings are not usually described in terms of acres but as 'rod-widths' or strips of one rod in width. The statute rod, pole or perch was 16½ ft, but locally 18 or 20 ft were more common lengths. It is quite clear that everyone who held a bovate of land in the 'Infields' of Snainton had the right to one 'rod-width' in each of the furlongs of the 'Outfields'.

Much more research will be needed before one can re-create a detailed image of the Yorkshire countryside as it existed in Norman times. Even so, contemporary access to first-rate evidence far exceeds that of a generation ago. Evidence in the Dales and, perhaps, on the Moors too, suggests the frequent existence of older hunting forest on the high ground, while thorn and scrub had spread across the land of wasted vills in the valleys, though other farmland had remained in use or was being recolonised. There was a basic division of tenure and land organisation which involved the communities tied to lay manors, the royal hunting reserves, which were set to expand, and the estates associated with the new monastic communities. The southern section of the Vale of York seems to have consisted of a complex mixture of farmland and waste, with field patterns slowly forming as assarting bit back into the scrub and woodland cover; two-field systems were numerous in the north. There is clear evidence of a planned and detailed reorganisation of the farmland in the Wolds and Holderness. The accumulating evidence suggests that a framework of long strips was established in middle Saxon times, that this pattern may have survived the Norman period and, in places, have been reorganised to produce more 'conventional' open-field systems in the centuries that followed.

REFERENCES

D. Austin, 'Medieval settlement in the North-East of England – retrospect, summary and prospect', in B. E. Vyner (ed.), *Medieval Rural Settlement in North-East England* (Durham: 1990), pp. 141–50.

T. A. M. Bishop, 'The Norman settlement in Yorkshire', in R. W. Hunt *et al.* (eds), *Studies in Medieval History Presented to Frederick Maurice Powicke* (Oxford: 1948).

F. W. Brooks, 'Domesday Book and the East Riding', *East Yorkshire Local History Series*, 21 (Beverley: 1986).

M. Chibnall, *The Ecclesiastical History of Orderic Vitalis*, 6 vols (Oxford: 1969–80).

H. C. Darby and I. S. Maxwell (eds), *Domesday Geography of Northern England* (Cambridge: 1977).

E. H. Freeman, *History of the Norman Conquest of England*, 2nd edn, 4 vols (Oxford: 1870–6).

J. A. Giles (ed.), *William of Malmesbury's Chronicle of the Kings of England* (London: 1847).

D. Hall, *Medieval Fields* (Aylesbury: 1982).

B. J. D. Harrison and B. K. Roberts, 'The medieval landscape', in D. A. Spratt and B. J. D. Harrison (eds), *The North York Moors*, (London: 1989), pp. 72–112.

M. Harvey, 'Regular field and tenurial arrangements in Holderness, Yorkshire', *Journal of Historical Geography*, 6 (1980), pp. 3–16.

M. Harvey, 'Planned field systems in eastern Yorkshire: some thoughts on their origin', *Agricultural History Review*, 31 (1983), pp. 91–103.

M. Harvey, 'The development of open fields in the central Vale of York: a reconsideration', *Geografiska Annaler*, 67B (1985), pp. 35–44.

S. P. J. Harvey, 'Domesday Book and its predecessors', *English Historical Review*, 86 (1971), pp. 753–73.

D. Hey, *Yorkshire from AD1000* (Longman: 1986).

J. C. Holt (ed.), *Domesday Studies* (Woodbridge: 1987).

D. Hooke, 'Early forms of open-field agriculture in England', *Geografiska Annaler*, 70B (1988), pp. 123–130.

W. E. Kapelle, *The Norman Conquest of the North: The Region and its Transformation 1000–1135* (London: 1979).

W. Matzat, 'Long strip field layouts and their later subdivisions', *Geografiska Annaler* 70B (1988), pp.133–47.

D. M. Palliser, 'Domesday Book and the "Harrying of the North"', *Northern History*, 29 (1993), pp. 1–23.

B. K. Roberts, 'Norman village plantations and long strip fields in Northern England', *Geografiska Annaler*, 70B (1988), pp. 169–77.

T. Rowley (ed.), *The Origins of Open-Field Agriculture* (London: 1981).

J. A. Sheppard, 'Pre-enclosure field and settlement pattern in an English township: Wheldrake, near York', *Geografiska Annaler*, 48B (1966), pp. 59–77.

J. A. Sheppard, 'Field systems of Yorkshire', in A. Baker and R. Butlin (eds), *Studies of Field Systems in the British Isles* (Cambridge: 1973), pp. 145–87

J. A. Sheppard, 'Pre-Conquest Yorkshire: fiscal carucates as an index of land exploitation', *Transactions of the Institute of British Geographers*, 65 (1975), pp. 67–78.

W. E. Wightman, 'The significance of "waste" in the Yorkshire Domesday', *Northern History*, 10 (1975), pp. 55–71.

CHAPTER TEN

Villages and Hamlets

VILLAGES WILL HAVE EXISTED in the country-sides of Yorkshire since at least Neolithic times, though throughout prehistory and the first few centuries of the historical era they were a subordinate facet of a settlement pattern that was dominated by hamlets and farmsteads. Dispersed forms of settlement still dominate the uplands, although nucleation of settlement is strongly developed in the vales. The evidence from some other parts of England shows that villages began to gel during the middle Saxon period, at about the same time that open-field farming systems were introduced and hundreds of field churches were established. Thus the nucleation of settlement appears to have been a crucial aspect of the so-called 'middle Saxon shuffle'. Hitherto, villages had been greatly outnumbered by farmsteads and hamlets and all settlements had tended to be rather short-lived. In the course of the changes, the ancient rural pattern of small, irregular fields, little woods, winding lanes and scattered settlements was replaced in many settings by one of villages, with dwellings clustered around the church and lord's house, which were located in the centre of the two or three great open fields. This process of nucleation and reorganisation, which probably began around the ninth century, continued to produce new villages and field systems during the earlier medieval centuries.

This being so, one would not expect many of the existing villages of Yorkshire to have pedigrees which extend back beyond late Saxon or Viking times, though, actually, the history of any particular village could be very complicated. One cannot excavate a living village to discover its origins and early layout, but such excavations have been attempted in villages that were deserted in medieval or later times. Of these excavations, by far the most important is the one that began at Wharram Percy in the Yorkshire Wolds in 1950 and continued until 1990. The detailed evidence from Wharram shows just how complicated the ancestry of a village can be. A Neolithic stone axe imported from the Langdale factory in Cumbria, discovered nearby, was probably a relic from the original clearance of the area; as the countryside filled up in the later part of the Bronze Age or early in the Iron Age, the lands here were partitioned by linear earthworks. Iron Age pottery indicating habitation around the village site dated back as far as the seventh century BC. In late Iron Age times, two farms existed on what later became the site of the medieval village. They were identified as 'ladder' settlements, which seem to have been common in East Yorkshire, so called because the form, with rectangular enclosures running off from the margins of a road, resembles a ladder in plan. Such ladder patterns lay on either side of the sunken road, with farmsteads dotted at intervals along the roadside. By the middle of the Roman era, the two farms had increased to at least five. The farms were strung out in a loose grouping about 200 metres apart. They may have been associated with a Roman villa, for the footings of what may have been its corn-drying furnace were found beneath the North Manor of the medieval village. This would have been one of a cluster of villas in this locality, either the centres of estates created to supply the population of Roman Malton with farm produce, or else the county mansions of Malton-based officials.

The nature of the links between the village and the earlier settlement on the site is uncertain. The Romano-British settlements were not necessarily directly ancestral to Wharram Percy, though when

The recently excavated and restored mill pond at Wharram Percy; beyond are the ruins of the church, which long outlived the deserted village

the medieval village took shape as a planned village, with house plots set around the ribbon of green which carried the through-road, its form was determined by the pattern of prehistoric ditches and Roman field boundaries; there is a possibility that the loose Romano-British settlement could have governed the layout of the western row of house plots. The fate of the Romano-British settlements after the collapse of the Roman economy is uncertain. The first evidence of Saxon settlement was provided by the discovery of two sunken floored buildings or *Grubenhauser* of the sixth century. The Wharram site has yielded Saxon pottery of various dates, some of it quite early; an eighth-century smithing hearth, a fragment of a cross of about 800, and a late Saxon mill-dam have also been found. It appears that, as yet, there was no compact village, but a loose scatter of farmsteads and dwellings similar to the earlier patterns. Until very recently

it was thought that the Scandinavian invasion and settlement had provided the change and disruption that instigated the reorganisation of rural life at Wharram. This would have involved the creation of the vast fields of long strips and the resettlement of the population in a purpose-built planned village. Certainly, the overwhelming predominance of Scandinavian place-names in the area seemed to argue that it fell into the hands of the new foreign aristocracy. In 1984, however, evidence of Scandinavian-period occupation was found in one of the village crofts, in a situation which suggested that the formation of a compact village had not been accomplished by this time, so the question of whether the village took shape

The deserted medieval village of Wharram Percy. The ruined and roofless church stands just to the right of centre. Melting snow outlines the holloways and house sites of the village; traces of the ridge and furrow ploughland are apparent to the upper and lower left of the photograph. (Anthony Crawshaw)

in the years before or after the Norman conquest remained open. In 1990, Beresford and Hurst, who had guided the long series of excavations at Wharram Percy, wote that:

> by the seventh and eighth centuries settlement at Wharram Percy seems to have revived to the extent that there were at least six main areas of settlement, a similar number to that in the Roman period ... Five of these areas were of high status with evidence for metalworking, coins and imported pottery. The general picture is similar to that in the Roman period, with farms spaced irregularly in the landscape. The

exact nature of the site is unclear but there is a distinct possibility that Wharram Percy may form one of the small family monastic sites that are known to have existed in the middle Saxon period.

Even after decades of painstaking excavation, however, the birth of the nucleated settlement could still not be determined:

> From the evidence elsewhere in England the change from scattered settlement to villages with compact groups of houses occurs during the late Saxon, Scandinavian or early medieval periods, that is in all periods between the ninth and twelfth centuries. It is still unclear when it happened at Wharram and whether the planned village was laid out at the time when the houses became grouped together, or whether the unsettled parts of the village gradually became filled in and the village was subsequently replanned.

The date of Wharram's appearance as a planned village remains to be confirmed, but the two main opportunities for a reshaping of settlement will have occurred with the Scandinavian domination of communities in the Wolds during the tenth century and with the dislocation caused by the Norman Harrying.

The example of Wharram reveals both the complexity of village origins in Yorkshire, and the way in which villages embody patterns of life which have been evolving for thousands of years. Other important excavations in Yorkshire have also explored the complexities surrounding the birth of village England. A project in the parish of Heslerton, begun in 1975, explored the archaeology of a section of landscape running from the chalk of the Wolds down into the Vale of Pickering. Evidence of settlement in the late Neolithic period was found here, together with a settlement that was established on the edge of the marshes which covered the heart of the Vale of Pickering in the early Iron Age. It expanded in a straggling form and persisted throughout the Roman era; as the climate deteriorated and the water-table rose in the last phases of Roman rule, so the site became increasingly wet. It was at about this time that new settlers arrived to join the indigenous people. They may have been Anglians from southern Denmark, and they built structures with distinctive scooped-out floors. Powlesland identifies these not as weaving sheds, the conventional interpretation, but as buildings for general storage, particularly the storage of grain, which could have had floors suspended over the pits to improve the air circulation. By about 450, the waterlogging caused the settlement to be abandoned, and a new, streamside site on firmer ground about half a mile (²/₃km) to the south was chosen. Substantial timber halls and *Grubenhauser* were erected. The cemetery of the pagan settlement lay 365 metres away, amongst a complex of Bronze Age barrows, and a ritual horse burial was found amongst the burials and cremations. As at Wharram, a locality favoured by intermittent settlement in the earlier prehistoric phases went on to attract settlement in the Iron Age and Roman periods which, by different stages, then graduated into a village during the Dark Ages.

At Crayke, on the edge of the Vale of York and the North York Moors, on an outlier of the Howardian Hills, excavations in the pre-fourteenth-century cemetery in 1983 suggested a settlement regime which ran from the Roman period to the present. A monastery was founded there in the seventh century by St Cuthbert, Bishop of Lindisfarne, on land granted by King Ecgfrith of Northumbria; he established a community of monks, under an abbot, in a locality which, at that time, was probably wooded. The settlement sequence established by the excavations began with early and late Roman settlement, separated by a gap in occupation during the middle Roman period; the hill might have been the site of a stone-built villa with a hypocaust. Then there was the establishment of a Christian cemetery, probably the cemetery associated with Cuthbert's monastery. Some time after 883, the monastery was abandoned and the landscape was reorganised, giving rise to a village with a triangular green and an open-field system; its North Field was set out across part of the old cemetery.

It seems certain that villages were beginning to take shape in Yorkshire in the centuries preceding the Norman conquest, when farming patterns were being reorganised and peasants migrated from their farmsteads and hamlets into the new, compact settlements; the building of village churches and manor houses as central places in the countryside accompanied and strengthened this centralising process. Even so, as Faull has shown: 'As matters stand at present, it will be many years before we can hope to be able to draw a map of late Anglo-Saxon settlement patterns in Yorkshire.' Whether many of the villages emerging at this time were set out to a particular plan, around the church and manor, or grew in a more organic or haphazard way is still debated. The majority of experts consider that scores of settlements were wasted in the Harrying and were subsequently replaced by villages which were very precisely planned. (There is no unanimity on this, and Austin wrote: 'I would point to the hoary old chestnut that village regularity came about after the Harrying of the North in the later eleventh century. There is no evidence to support this notion of process nor the dating. In fact all the archaeological evidence we have suggests a broad range of processes and dates from the 9th to the 13th centuries (and beyond), while documentary evidence tells us nothing except to reaffirm the immense complexity.' It is true that village formation in Yorkshire continued throughout much of the Middle Ages, particularly in the more sparsely populated areas, and these new villages could be either planned or organic in form.)

A high proportion of the villages in the Vale of York still display traces of their planned origins. This regulation is evident in layouts where the dwellings were set out beside the road or roadside green in neat rows, and where each dwelling stood in a long narrow plot or 'toft' which ran back from the roadside or green, with the tofts terminating at a continuous boundary, sometimes

marked by a back lane, as at Wetwang, Appleton-le-Moors, Middleton, and many other places. The tofts were often all of the same size, though later amalgamations may blur the pattern. Frequently, two rows of dwellings and tofts would face each other across the road, or road and green, as at Huby, but usually there would just be one row, as at Gate Helmsley, and sometimes three rows, arranged around a triangular green, as at Nun Monkton. When such villages were set out, their form and dimensions might have been determined with great precision. Sheppard has noted that about half the planned villages she studied had the lengths of their frontages determined exactly by the data recorded in the Domesday tax assessment – namely, two perches of frontage per bovate. Their dimensions also seem to imply the use of a measuring rod by the surveyors who set out the sites, and there was a marked tendency for multiples of 5.5–6 metres to emerge. Since the Domesday tax assessment was superseded early in the thirteenth century, these planned villages are presumed to have appeared between 1086

A corner of the planned medieval village of Arncliffe, the buildings still arranged around the central rectangular green

and then; it is not easy to resist the argument that they were built in the aftermath of the Harrying, when lords were seeking to revive their wasted estates. The villages could have been used to accommodate local survivors and other peasants drafted in from the west, but they could also have attracted new settlers lured by the offers of viable holdings on good land.

The planned villages seem to have appeared in vills which lacked demesne land (land worked directly for the lord) and which were held by one of Yorkshire's 26 tenants-in-chief. It is suggested that these powerful men were seeking to stimulate recovery by attracting tenants to the most devastated areas, offering to accept rent payments rather than onerous obligations of bondage and service. There is also the suggestion that local survivors were sometimes accommodated in remaining, unplanned villages – and made the subject of harsh manorial services. The relationship between the planned settlements (presumably) of the post-Harrying years and landscape features inherited from the preceding periods – like Saxon churches – would reward investigation; some early churches seem to be eccentrically located in regulated villages. Appleton-le-Moors, near Pickering, is invariably quoted as a prime example of village planning, but there are scores of other places where the evidence of medieval regulation is preserved in the layout of a living

village. Appleton may not be a product of the Harrying; it seems to have come from an amalgamation of the original Appleton and nearby Balskerby, a lost village, perhaps in the twelfth century. Village planning was certainly the norm in the Vale of Pickering in early medieval times. Pamela Allerston studied the maps of 29 villages in the Pickering area and recognised evidence of planning in 18 cases, but then the archaeologist Christopher Taylor examined the villages and their adjacent earthworks on the ground and thought that 23 or, perhaps, 25 of them showed traces of medieval planning. Another well-preserved example of a planned village is Cold Kirby, on the moors near Helmsley, though here, as in some other northern villages, the rectangular green was enclosed early in the nineteenth century and incorporated as front gardens.

Not all village regulation was a consequence of the Harrying, and planned villages are not confined to the lowlands. One well-known and very well-preserved example is Old Byland, built to house a community evicted by the monks of nearby Byland in the 1140s; another is East Witton, probably developed by the monks of neighbouring Jervaulx around 1300, with dwellings set out on either side of an elongated market green; Arncliffe is a planned example from deep in the dales; and Newton-on-Rawcliffe, with its neat green and pond, is another from the edge of the North York Moors. Wharram Percy was planned with two long rows of dwellings and one much shorter row lining the sides of the green which bordered the village road and had the shape of a very elongated triangle. Each house plot was 18 metres wide, the width of two field strips, although the village form was also tailored to the existing framework of Roman and earlier boundaries. The village layout here was frozen in time after the desertion of the site around 1500. Had Wharram continued to flourish, its carefully

planned origins might have been obscured by later development. Nearby Wharram le Street has acquired a T-shaped layout, although it began as a planned village of two-house rows, and another neighbour, North Grimston, completely outgrew its regulated format in the course of the last century.

In some places, the clear evidence of early medieval regulation has endured in the village landscape for almost a millennium; in other cases, it has been severely blurred by later growth or shrinkage. There are also Yorkshire villages which were never laid out to a set pattern. Some unplanned villages coalesced very gradually from the merging of separate clusters of dwellings and farmsteads; hundreds of years later, it is still possible to recognise the old nuclei, sometimes represented by churches or different greens. Such settlements were designated as 'polyfocal' villages by Taylor. Some are very old and may result from the gradual blending of hamlets that had grown up around particular settlement nuclei, such as an early church, a manor or a small monastic house. Others can be relatively recent; Darley, in Nidderdale, resulted from the merging of a series

---

The medieval village was very largely dependent on the produce of its surrounding fields. Here, at Burnsall, in Wharfedale, some of the medieval patterns of cultivation are evident in the adjacent fields

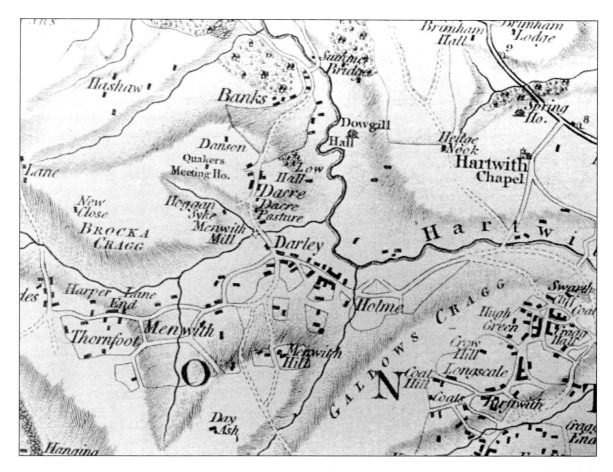

of hamlets and mill-site clusters which were strung like beads along a lane leading up the dale to Pateley Bridge. Growth associated with an early stage in the Industrial Revolution, represented in Nidderdale by water-powered linen- and twine-milling, provided the stimulus for this example of 'late nucleation'.

It is not clear how much influence this tendency for villages to gel from the earlier farmstead and hamlet patterns had exerted in the dales and upland country before the Norman conquest; there were probably a few, but not many, cohesive villages in the dales before the Harrying. Afterwards, however, the enthusiasm for hunting forests and the carving out of great monastic estates would have restricted the opportunities for village creation, though they did lead to the establishment of a small number of villages, like Bainbridge and Buckden, which were specifically founded to accommodate foresters. Raistrick wrote that:

Many of the elements in the village landscape of the Yorkshire Dales did not gel until the Industrial Age. This is a part of Nidderdale as mapped by Jeffrey at the start of the last quarter of the eighteenth century, when industry was beginning to exert an effect upon the landscape. Summerbridge, Dacre Banks and Dacre would emerge as cohesive villages, though the author's village, near the centre of the right-hand margin, did not appear until the nineteenth century. Darley existed as a 'polyfocal' village, with dwellings set around a number of distinct foci. At this time, dispersed farmsteads and loose hamlets – like the ones in the lower right quadrant – were still the most characteristic features of the countrysides of the dales

The village of Bainbridge was built as the chief lodge of the forest [of Wensleydale], and a *quo warranto* document of 1227 AD calls upon Ranulph son of Robert to answer Ranulph, Earl of Chester and Lincoln by what warrant he made towns and raised edifices in the Earl's forest of Wensleydale. He answered that the town of Bainbridge was of the ancestors of the same Ranulph by service of keeping that forest so that they should have there 12 foresters each to have a house and 9 acres of ground . . .etc.

The process of village formation continued in the Dales until after the close of the medieval period; in Upper Nidderdale, for example, the villages and hamlets, like Ramsgill, Middlesmoor, Lofthouse and Bouthwaite, developed gradually from monastic granges in late medieval and later times.

In the course of the Middle Ages and the centuries that followed, the design of some villages proved remarkably stable, making it easy to recognise their original layouts. Roberts (1990) comments that: 'It is, of course, true to say that no village truly survives from the twelfth or thirteenth centuries – although odd things can and do happen in darkest Yorkshire! . . . Just as the twelfth or thirteenth century church may survive, even when altered, adapted and even substantially rebuilt by later generations, so may elements of a village plan – although medieval society, economy, technology and culture have long gone.' As will be seen in Chapter 12 below, there were many other villages which shrank or disappeared altogether during the Middle Ages, and some which adopted new alignments or even moved across the map. Killinghall, near Harrogate, belongs to this last category; in the eighteenth century it existed as a string of farmsteads along the edge of a large common. Following the enclosure of the common late in that century and the turnpiking of two tracks which ran across the common, new settlement was attracted, leaving the farmsteads of the older village stranded across the fields. Just across the Nidd lies Ripley, another shifted village, where the cemetery of the original village was recently excavated by Cale. Here it seems likely that the shift took place after the Ingilby family gained control of the manor in the fourteenth century. Sir Thomas de Ingilby was granted a charter for a market and fair in 1357, which may have prompted him to shift the village about 365

metres to a site astride a Roman road, which may have brought more custom to his new market. The old church was abandoned and a new one was built.

It was during the Middle Ages that the village patterns of Yorkshire became firmly established. The areas that were richest at the time of the Norman conquest – the Vales, the Wolds and Holderness – became village country, with relatively modest numbers of hamlets and scattered farmsteads, although some existing upland farmsteads in the Wolds stand on sites that formerly supported hamlets. The Dales and the Pennine foothills, meanwhile, developed a mixture of villages, mainly lying in the valleys, and a heavy scattering of hamlets and farmsteads.

The eighteenth and nineteenth centuries contributed a few additions to the landscape, in

The medieval market cross behind the village stocks at Ripley. The mock-Elizabethan house seen beyond the stocks reflects the wholesale rebuilding of the village by its owners in about 1822

the form of planned estate or industrial villages. Harewood, built outside the park gates of Harewood House, near Leeds, is probably the best-known example. Less well-known today is the village of Boston Spa, near Wetherby, built in open-field land in 1753 to exploit a mineral spring and, thereby, profit from the rising fashion for spas. Fulneck, near Pudsey, was the first of several villages in Britain to be built by the Moravians, a German Protestant sect, and was based on the German village of Herrnhut. It was completed in 1748, on a difficult hillside shelf, as a new and idyllic home for the brethren. Copley and Akroydon, near Halifax, were built in the middle of the nineteenth century by Colonel Edward Akroyd, MP, to provide improved accommodation for millworkers; similar ideals at Saltaire produced a town of more than 800 houses, the creation of Sir Titus Salt. Most of Baldersby, near Thirsk, was created by the Victorian architect William Butterfield for Viscount Downe in the 1850s.

Although considerable attention has been lavished on villages, research relating to hamlets has been limited; in Yorkshire, however, interesting work has been done by McDonnell. Expansion into the wastes during the medieval period was associated with a friendly climate and the pressures of an expanding population, but the era of colonisation was followed by one in which a declining population retreated from the margins. In Bilsdale, on the North York Moors, during the late thirteenth and early fourteenth centuries, landlords seeking to reclaim and intake the marginal uplands and exploit their timber resources offered favourable freehold or leasehold terms to tenants prepared to assart the waste at the heart of the dale. Foresters, carpenters, cowmen and other volunteers took up the challenge. McDonnell has described how this resulted in the establishment of at least four hamlets, each consisting of around six holdings, with the tofts or house plots grouped together for mutual protection and with small areas of common ploughland. In most cases, however, the life of the settlements was merely two generations: Scottish raids threatened; the canons of Kirkham may have called in their leases; and, most likely, as the climate deteriorated and the economy weakened in the first third of the fourteenth century, the tenants chose to abandon their hamlets. Urra, with good pastureland, preserved its pattern of settlement into the seventeenth century, but the other hamlets contracted to single farmsteads. The Cistercian granges in the south of the dale experienced similar changes, so that what had been a settlement pattern of small clusters became one of solitary farms. McDonnell points out that hamlets cannot simply be characterised as failed villages; some might result from the subdivision of a farmstead, a process which can ultimately lead to the appearance of a village. There are several villages in the Dales, like Kilnsey in Wharfedale, which have evolved from granges, though in the uplands 'development tended to stop short at the hamlet stage through paucity of resources'. In Swaledale, small nucleated settlements have derived from monastic cattle farms or vaccaries; originally worked by lay brothers attached to the Cistercian abbeys, the vaccaries began to be let to laymen during the fourteenth century, when grangers from the abbey community were replaced by paid managers or lay grangers. According to McDonnell, many upland hamlets, like those examined in Upper Swaledale, in other central Pennine Dales and Craven, showed signs of 'several facets of settlement evolution which are usually obscured in more developed villages; we can see the bare bones of the original layout and appreciate the reasons for it'. Some of the vaccary settlements developed into genuine villages, like Gunnerside and Healaugh, while others, like Kearton, Smarber and Birkdale, shrank, expanded with mining, then dwindled again. Others still are hamlets, fossilised at a stage when their stock-farm origins could be preserved, with rather small central greens of about half an acre which derive from the fold-yards where the cattle were penned: 'There seems no reason to doubt that these modest dimensions quite accurately preserve the lineaments of the original vaccary fold-yard and its surrounding buildings.'

REFERENCES

K. A. Adams, 'Monastery and village at Crayke, N. Yorkshire', *Yorkshire Archaeological Journal*, 61 (1990), pp. 29–50.

P. Allerston, 'English village development: findings from the Pickering district', *Transactions of the Institute of British Geographers*, 51 (1970).

M. Aston, D. Austin and C. Dyer (eds), *The Rural Settlements of Medieval England: Studies Dedicated to Maurice Beresford and John Hurst* (Oxford: 1989).

D. Austin, 'Medieval settlement in the North-East of England – retrospect, summary and prospect' in B. E. Vyner (ed.), *Medieval Rural Settlement in North-East England* (Durham: 1990), pp. 141–50.

M. Beresford and J. Hurst, *Wharram Percy: Deserted Medieval Village* (London: 1990).

K. Cale, *Yorkshire Archaeological Journal*, forthcoming.

M. Faull, 'Late Anglo-Saxon settlement patterns in Yorkshire', in M. Faull (ed.), *Studies in Late Anglo-Saxon Settlement* (Oxford: 1984), pp. 129–42.

R. Fieldhouse and B. Jennings, *History of Richmond and Swaledale* (Chichester: 1978).

J. McDonnell, 'Medieval assarting hamlets in Bilsdale, North-East Yorkshire' *Northern History*, 22 (1986), pp. 269–79.

J. McDonnell, 'Upland Pennine hamlets', *Northern History*, 26 (1990), pp. 20–39.

R. Muir, 'The villages of Nidderdale', *Landscape History*, 20 (forthcoming)..

D. Powlesland, 'West Heslerton', *Current Archaeology*, 76 (1981), pp. 142–4.

D. Powlesland, with C. Haughton and J. Hanson, 'Excavations at Heslerton, North Yorkshire 1978–82', *Archaeological Journal*, 143 (1986), pp. 53–173.

A. Raistrick, *Arthur Raistrick's Yorkshire Dales* (Clapham: 1991).

B. K. Roberts, *The Making of the English Village* (Harlow: 1987).

B. K. Roberts, 'Back lanes and tofts, distribution maps and time, medieval nucleated settlement in the North of England', in B. E. Vyner (ed.), *Medieval Rural Settlement in North-East England* (Durham: 1990), pp. 107–26.

B. K. Roberts, 'Dating villages: theory and practice', *Landscape History*, 14 (1992), pp. 19–30.

B. K. Roberts, *Landscapes of Settlement* (London: 1996).

J. A. Sheppard, 'Pre-enclosure field and settlement patterns in an English township – Wheldrake, near York', *Geografiska Annaler*, 48B (1966), pp. 59–77.

J. A. Sheppard, 'Metrological analysis of regular village plans in Yorkshire', *Agricultural History Review*, 22 (1974), pp. 118–35.

J. A. Sheppard, 'Medieval village planning in Northern England: some evidence from Yorkshire', *Journal of Historical Geography*, 2 (1976,) pp. 3–20.

C. C. Taylor, 'Aspects of village mobility in medieval and later times', in S. Limbrey and J. G. Evans (eds), *The Effect of Man on the Lowland Zone*, Council for British Archaeology Report 21 (1978), pp. 126–34.

C. C. Taylor, 'Medieval market grants and village morphology', *Landscape History*, 4 (1982), pp. 21–8.

C. C. Taylor, *Village and Farmstead* (London: 1983).

C. C. Taylor, 'Medieval rural settlement: changing perceptions', *Landscape History*, 14 (1992), pp. 5–64.

C. C. Taylor and R. Muir, *Visions of the Past* (London: 1983).

CHAPTER ELEVEN

Yorkshire Monasteries

YORKSHIRE IS NOTED for its monastic heritage and many would argue that, in Fountains and Rievaulx, Yorkshire possesses the two finest medieval abbey ruins in Britain. When the total collection of monastic relics, including places like Byland, Bolton Priory and Whitby, is considered, there is little doubt that the assemblage is unsurpassed by that of any other British region. The tranquillity associated with many of these places is misleading; today we see the monasteries as ruins, gaunt, romantic and soulful, but it is important to remember that, in their day, they were vibrant places which hummed with activity. Fountains came close to resembling an industrial site, with its stream polluted and a constant pulse of traffic moving to and from its far-flung holdings. Monasteries were the organisational centres of vast estates, the focuses for scholarship and the seats of great power, with incredible wealth fossilised in their buildings and fittings – as well as being the homes of communities which could

be considerably larger than the populations of nearby villages.

Although several monastic orders were represented there, Yorkshire was primarily Cistercian territory; countrysides still preserve the marks of Cistercian control and the circumstances of Norman Yorkshire proved particularly conducive to Cistercian expansion. However, the Cistercians were by no means the first to establish monastic houses in Yorkshire; in the centuries before the conquest, the Benedictine order was paramount throughout England. The Benedictines in Yorkshire suffered the full violence of the Viking raids. Some of their houses were so devastated that they never revived and one is still unable to identify the entire complement of their early foundations. Burton writes that:

The medieval monasteries of Yorkshire exist as gaunt and soulful ruins; this is Easby

The 11th and 12th centuries witnessed a great intellectual questioning of the nature of monastic life and from this debate grew a number of new religious orders. By the time monasticism returned to the north of England there was a great variety of forms of religious observance and of monastic orders which founders of religious houses could choose to patronize. In the period up to the 10th and 11th centuries European monasticism was dominated by Benedictine monasticism, by those houses which based their way of life on the monastic rule composed by St Benedict of Nursia in the 6th century for his community of Monte Cassino in Italy.

Yorkshire's conversion – or reconversion – to Christianity and the establishment of the first generation of minster churches predated the emergence of a formalised Benedictine order here. A key figure in this early process was Benedict Biscop, who encountered continental monastic practices during his travels and studies in Italy and Gaul. He made six visits to Rome, bringing back books which would be studied by Bede. Wilfrid, who grew up in the Celtic foundation on Lindisfarne, drew upon his experience abroad – he visited Rome before he was 20 – to introduce a continental form of observance at his monastery at Ripon in 661; and he presided over the rebuilding of the Minster at York, where, it was said, the roofless church was repaired, its windows glazed and its walls limewashed. He was also said to have settled monks in places abandoned by the British clergy during the Saxon invasion: 'round Ribble and Yeadon, the region of Dent, Catlow and other places'. Meanwhile, Benedict Biscop was active in the lands further north, establishing churches and monastic communities at Monkwearmouth in 674 and Jarrow in 681. The outlines of the Benedictine order were established by Benedict, who was born about 480 and who established the abbey at Monte Casino in Italy; the rule of St Benedict influenced the organisation of the communities established by Wilfrid and Benedict Biscop in the North and, in due course, York emerged as an international theological centre which was influential in missionary work in Germany. Even so, the Benedictine

rule was not fully established as the model for monastic organisation until 817.

The three wealthiest Benedictine establishments in medieval Yorkshire were those at Selby, at St Mary's, York, and at Whitby. Selby was a post-conquest foundation of about 1070, and the abbey church, of the Norman cruciform-plan type, survived the traumas of the Dissolution of the monasteries. Even so, it has had an eventful history; the distorted arches at the end of the nave result from the sinking of the tower foundations during the building work. The tower collapsed in 1690, bringing down the south transept, and the roof and rebuilt tower were burnt in 1906. The founder of Selby was another Benedict, a monk from Auxerre in France, who had a personal mission to revive monasticism in the North of England. Legend tells that he was inspired by a vision to erect a cross and a hermitage on a spot beside the lower Ouse, and that this cross was seen by William the Conqueror's sheriff, Ralph FitzBaldrick. The king was impressed and made a grant of land upon which Benedict of Auxerre established his monastery. Such high ideals were not always upheld by his successor, however, and in 1281 Abbot Thomas de Whalley was deposed by the Archbishop of York for the mismanagement of his estates and dissolute conduct – he had two mistresses and several children. St Mary's, York, was another post-conquest foundation. The monks at Whitby had an unhappy relationship with the great de Percy family, and this resulted in the departure of Abbot Stephen with a dozen of his monks. They were given 4 acres of land in York, near St Olave's church, by Alan, Earl of Richmond, on which their new abbey was founded. Generous endowments were received from King William Rufus, others followed, and the abbey gained control of some 44 churches in Yorkshire. The original foundation was burned in 1137 and the surviving buildings date from 1270; St Mary's is unusual in having its own defensive wall.

The Benedictine abbey at Whitby had a much longer pedigree. The original foundation dated back to 657, when King Oswy of Northumbria founded the monastery to celebrate his victory over Penda, the pagan King of Mercia, at Winwaed some twenty years earlier (originally, he promised to found 12 monasteries if the battle went in his favour). St Hilda, baptised by Paulinus at the age of 13, was brought from Hartlepool to serve as abbess over the double monastery (i.e. for men and women) at Whitby; the reputation of the community grew so quickly that Whitby was chosen as the site for the famous synod of 663. In 867, however, the monastery perished in a Danish raid. In 1078 the Norman monk, Reinfrid of Evesham, refounded the monastery. Like Benedict of Auxerre, he had moved to the North with a mission to re-establish the monastic tradition which had been so devastated by the Viking raids, and he arrived at Whitby with his companions after working at Newcastle and Jarrow. William de Percy granted the site for the new priory, which was elevated to the status of an abbey early in the twelfth century. Around 1153, the town and abbey were plundered by the forces of the King of Norway, yet, at the end of this century, the foundation was prospering and supported about 40 monks. Part of the abbey church survives, unroofed and over 91 metres in length, displaying architecture of the thirteenth and fourteenth centuries. The north wall of the north transept is the best preserved and most spectacular feature, with three tiers of lancet windows crowned by a rose-windowed gable and flanked by turrets.

The post-conquest advance of monasticism in Yorkshire might have been spearheaded by the Benedictines, but the Cluniacs were the first to enjoy royal favour after the conquest. Originating at Cluny in Burgundy and emerging as a distinct order in the eleventh century, the Cluniacs had an almost obsessive enthusiasm for liturgy and ceremony which appealed to the Norman kings. Some 36 Cluniac priories were established in England, including one at Pontefract, but, in the event, the introspection of the order, its repatriation of revenues to France and the practice of founding priories which were subordinate to the Abbot of Cluny eventually deprived it of favour and momentum. The priory of St Mary Magdalene at Monk Bretton, near Barnsley, was colonised from Pontefract in 1154. However, its charter was ambiguous and a prolonged dispute reigned over the right to appoint the prior, resulting on occasions in armed invasions from Pontefract and the refusal by the Monk Bretton community to accept an official Cluniac delegation. As a consequence,

the community left the Cluniac fold and became an independent Benedictine house in 1281. Perhaps the most interesting of the fragments to survive is the prior's lodging, converted into a private residence for one of the Earl of Shrewsbury's sons after the Dissolution.

It was the activities of monks of the Cistercian order which did so much to transform the face of Yorkshire, although their arrival was preceded by a period in the early twelfth century when the Augustinians had gained favour in the county. The flexibility of the Augustinian rule was a source of strength, allowing either a vigorous or a contemplative life. The history of monasticism reveals the periodic eruption of movements which reacted against what was seen as the laxity of contemporary practice and sought a return to a strict monastic code of conduct. In 1097, at Cistercium or Cîteaux in Burgundy, the Cistercian order was founded as an attempt to re-create the presumed austerity and purity of the first Benedictine communities. Thirty years passed before the Cistercians secured a foothold in England (at Waverley, in Surrey), but the arrival of the order in Yorkshire did not represent an expansion from Waverley. Whether or not the monks at Cîteaux were fully aware of it, Yorkshire was ideal Cistercian territory; there were extensive tracts of wild, thinly populated countryside, lords who were devout if harsh, and estates which had still to recover from the consequences of the Harrying. The Cistercians, meanwhile, had a rule which required the monks to minimise their contacts with the laity, and values which encouraged the colonisation of remote and difficult wildernesses.

In 1132, motives similar to those which had inspired the formation of the Cistercian order in Burgundy led a sub-prior, sacrist and 11 monks to leave St Mary's, York. They found a sympathiser in their archbishop, Thurstan, who was abused by the St Mary's community when he attempted to mediate in the dispute, and he was obliged to settle the dissidents on derelict land beside the River Skell. Here they camped and were said almost to have expired in the harshness of the ensuing winter, surviving until 1135, when they received an endowment from Hugh, a Dean of York, who resigned his deanship to join the community. In the meantime, they had made a successful

application to St Bernard, Abbot of Clairvaux, for enrolment into the Cistercian order. The abbey that they built, St Mary ad fontes, or 'at the waters', became anglicised and known as Fountains.

Fountains was not the most celebrated Cistercian abbey to be established in early medieval Yorkshire, neither was it the first. More famous was Rievaulx; the grant for this site in the Rye valley was made in 113I and Rievaulx was colonised directly from Clairvaux in 1132, after St Bernard had petitioned King Henry I and Arch-bishop Thurston. The expansion was remarkable; the foundation is said to have supported 300 monks and lay brothers within a decade of the commencement of work and, when the abbey was about 30 years old, it supported 140 monks

and 500–600 lay brothers. The third great Cistercian abbey in Yorkshire was Byland. It began as a Savignac foundation, but this order was absorbed by Cîteaux in 1147. The original colonists came from Furness Abbey in Cumbria in 1134, via Calder, where their vestigial foundations were plundered by the Scots; to Hood, near Thirsk, where the site proved too small; on to Byland, which proved to be uncomfortably close to Rievaulx; and, finally, to a marshy and overgrown site at Stocking in 1147. Work on the creation of a satisfactory and, eventually, imposing abbey took up much of the remainder of the century.

The best preserved of Yorkshire's Cistercian abbeys was not a member of this big trio, the 'three luminaries of the north', but Kirkstall,

Leeds. It was colonised from Fountains, via Barnoldswick, where the site proved to be poor and infested with bandits (a problem which also plagued the canons at Bridlington and Gainsborough and the monks at Whitby). The buildings, many of them intact to the level of the eaves, date from the late twelfth century; their survival through the medieval period with little alteration may reflect the community's insolvency in the thirteenth century, when there was much rebuilding elsewhere, and the onslaught of the Black Death in the century which followed. At Roche, near Maltby, in contrast, most of the buildings have gone, but the ground-plan is well-preserved. This is a granddaughter of Fountains, colonised from the daughter foundation at Newminster, in

Whitby Abbey was re-established in 1078, having been destroyed by Danes in 867–9. In the mid-twelfth century it was recorded that 'thieves and robbers, by day and night laid that holy place desolate. In like manner, pirates void of all compassion landed there, came and plundered the monastery'

Northumberland, in the late twelfth century. Approximately 20 Cistercian houses were established in Yorkshire; other notable members of the flock included Jervaulx, Meaux, near Hull, and Sawley. The foundation of Meaux Abbey by the Count of Aumale has been attributed to the cunning of a Cistercian monk called Adam. Burton (1989) relates it as follows:

St Mary's, York was one of the wealthiest of the Benedictine establishments in Yorkshire. It resulted from a breakaway from Whitby and gifts from Alan, Earl of Richmond, and from William II. The surviving buildings date from a rebuilding in about 1270

It was while Adam was at Vaudey [a daughter house of Fountains] supervising the construction of the monastic buildings that Aumale confided in him his anxiety concerning a vow which he had made (possibly at the time of the preaching of the second crusade in 1147) to go to the Holy Land. Advancing years and corpulence made him less willing to fulfil this vow. Adam evidently saw a golden opportunity to further the interests of the Cistercian order, and accordingly he suggested to Aumale that the foundation of another Cistercian house might adequately compensate for his failure to go to Jerusalem.

Far less influential than the Cistercians were the Premonstratensian canons, an order founded at Prémontré, near Laon, in 1151 to set an example of austerity and discipline to other canons. The impact of the order was modest: Bayham Abbey in East Sussex was its grandest foundation in England, though it did establish Easby Abbey, beside the Swale near Richmond, and Coverham Abbey in Wensleydale. As the order developed, Premonstratensians became more involved in contemplation than in missionary work, though they remained more interested than the Cistercians in pastoral work and established schools for local children in some churches. In contrast, the canons of St Augustine, who also sought to stiffen the conduct of the canons, had an enormous impact upon Yorkshire. They enjoyed considerable popularity in their day, controlling scores of churches and their valuable tithes, providing them with vicars and becoming much more involved in day-to-day life than the remote

St Mary at the Waters, better known as Fountains Abbey, is, for many people, the Cistercian abbey par excellence

Cluniacs and the reclusive Cistercians. The canons at Bolton, for example, served the parish churches at Skipton, Kildwick, Harewood, Broughton and Long Preston. Some of their priory churches were shared with lay parishioners, assisting their survival during the harsh times of the Dissolution. St Mary the Virgin, Bridlington, is a well-preserved example. The initial endowment of Bridlington was five churches and the canons subsequently gained control of seven more.

The first Augustinian priory was established at Colchester in 1103, and the canons soon found favour in the North. A foundation at Nostell, near Pontefract, was established in 1113, and Bridlington followed in 1115. Other notable houses appeared during the first half of the century at Kirkham, Guisborough, Warter, near Beverley, Drax, near Selby, and Newburgh, between Ripon and Malton. One of the most beautifully situated of Yorkshire's medieval monastic houses is the Augustinian priory of the Blessed Virgin Mary and St Cuthbert at Bolton-in-Wharfedale. Legend tells that its foundation commemorated the drowning of a young lord in the rushing waters of the nearby Strid, but the canons arrived here in 1154–5, after attempts to establish a daughter

house of Huntingdon Priory at Embsay, near Skipton, had proved unsatisfactory because of the poverty of the land. Bolton gained independence from Huntingdon in 1194, but remained a modest foundation, partly because of its vulnerability to Scottish raids. The priory church remains in use and displays excellent Decorated architecture – rather unusual at a Yorkshire monastic site, and a consequence of the destruction of the previous church during a particularly violent raid in 1320. Guisborough, now in Cleveland, was a much more prosperous foundation, though one which also boasted a reputation for piety. At the time of the foundation, some time between 1119 and 1124, Robert de Brus endowed it with about 10,000 acres of land. This wealth must have underlain the decision to undertake a rebuilding early in the thirteenth century, but a fire in 1289 and severe Scottish raiding undermined the finances of the priory in the fourteenth century. The main surviving feature is the east end of the

The famous cellarium *at Fountains dates from the late twelfth century. This portion was used for cellarage*

church, which dates to the end of the thirteenth century; there is little trace of the fortifications which were approved in 1344. Kirkham Priory, near Malton, was founded in about 1122 by the same Walter l'Espec, lord of Helmsley, who founded Rievaulx a decade later. The appeal of the Cistercian rule nearly led to the end of Kirkham as an Augustinian house during its early years; it remained of relatively modest size. The surviving architecture of the priory is fragmentary, the most notable feature being the remains of the gatehouse and lavatorium.

Yorkshire gained an incomparable collection of around 70 monasteries in the course of the medieval period. Benedictine nunneries existed at Nun Monkton, York and Wilberfoss; there were Cistercian nunneries at Esholt, Sinningthwaite,

Nun Appleton, Swine, Basedale and Hampole; a Cluniac nunnery at Arthington; the Carthusians had charter-houses at Mount Grace, on the North York Moors, and at Hull; and the Gilbertine canons had houses at York, Malton and Elleton. The Church was by far the most affluent institution in the land, and Yorkshire had a generous share of the 348 monasteries and 140 nunneries which England supported in the middle of the thirteenth century. While the relative fortunes of the houses fluctuated, the league of affluence in Yorkshire in the early thirteenth century was probably headed by St Mary's, York, Fountains, Guisborough, Selby and Rievaulx – although the interrupted character of the architecture in the splendid nave at Selby is thought to indicate a cash-flow problem. As well as controlling incredible wealth, the Church also wielded political power. The abbots of the three great Benedictine foundations had seats in the House of Lords, and a few, like those named and the fifteenth-century abbots of Fountains and Jervaulx, were mitred abbots and ranked with bishops. In addition, the abbots and priors were almost supreme within their own estates, dispensing a form of justice which was not always tempered with mercy. The estates of the greater abbeys could be vast; although Meaux is now a rather forgotten Cistercian foundation, it once controlled some 20,000 acres.

The monastic ideal had enjoyed enormous popularity in Yorkshire under the Normans. Burton (1987) notes that there had been no monasteries there at the time of the conquest, but, by the close of the twelfth century, there were more than 50 religious houses in Yorkshire. Enormous resources were needed to fund this new monastic community, and one of the most popular forms of endowment was the parish church. Many churches were possessed by wealthy laymen and the unpopularity of this arrangement encouraged the granting of churches to religious orders. Robert de Brus granted eight churches to his foundation at Guisborough; the canons at Drax controlled seven churches spread across Yorkshire and the Midlands; and in 1115 Count Stephen of

Aumale conveyed some sixteen churches to the Norman monastery at Aumale. Tithes could be expropriated without the transfer of responsibility for the churches concerned; St. Mary's, York, controlled the tithes of 30 parishes in this way. By the end of the twelfth century, 200 Yorkshire churches had been conveyed to monastic houses.

While Fountains and Rievaulx are remembered for their wool trade, they were also active in a range of other areas, like cattle farming, lead and iron-ore mining. In 1321 the modest priory at Bolton owned 1,579 ewes, 162 cows, 76 horses and 37 pigs. Yorkshire monasteries were at the forefront of the wool export trade, which was central to the English economy. Raistrick wrote that:

A typical list of some of these [Florentine] merchants has survived from the period 1280

This beautiful site in the Rye valley was colonised directly from Clairvaux in 1132; within a decade, Rievaulx may have supported some 300 monks and lay brothers. The cramped nature of the hillfoot side required the adoption of an unusual plan

to 1315 and gives a good idea of the scale of the trade. In one year the crop for the monasteries they dealt with in all the country was 3,291 sacks, each sack being 364 pounds weight. Of this nearly half was from Yorkshire monasteries and nearly half of this was from those of the Cistercians. In this list Fountains is put down as selling 76 sacks, that is 27,664 pounds.

Monastic agriculture tends to be regarded as being focused on the upland pastures of the Dales – and Fountains did control roughly a million acres in the uplands of Craven – but it was also important in East Yorkshire. Burton (1989) records that:

> The Cistercian monks of Meaux acquired pasture in Myton sufficient to graze 800 sheep, in Warter for 360, in Alverley (WR) and in Kirk Ella for 200 and in Moor Grange for 300. By the late 13th century Meaux's flocks in Holderness totalled 11,000 sheep and 1,000 other animals. Rievaulx had extensive pastures on the North Yorkshire Moors and in the East Riding at East Heslerton for 1,000 sheep, at Folkton for 1,000 sheep and at Humanby and Seaton Ross, and those of Malton pasture for 200 sheep at Newton.

The proportion of Yorkshire that was owned by the monasteries is not known, but Smith estimated that the Church held 27 per cent of freehold income from land in the territory. At the Dissolution, the monastic land was seized by the state and sold swiftly to finance wars against Scotland and France.

Lands bequeathed to the abbeys were reorganised and integrated into the monastic system of farming. Closer to the abbeys, the emphasis was placed on establishing arable farming wherever possible; on estates at some distance from the abbey site, granges worked by lay brothers and scattered subordinate lodges were characteristic of the Cistercian estates. Cistercian granges were established in 1134, under a statute of the Cistercian general chapter which required that they should be staffed by lay brethren, not monks, and should not be more than a day's journey from

the mother house. This was regularly flaunted, as in the case of the grange at Kilnsey, separated from Fountains by difficult country which only a trained athlete could have traversed in a day. One or two monks might be based in a grange to minister to the lay brethren and abbey tenants and to oversee activities, while a cellarer would arrive from the abbey from time to time to check accounts, collect rents and preside at courts. Vaccaries or cattle farms were characteristic of some areas of monastic farming, like Swaledale. The typical vaccary would handle a herd of 20–80 cattle and would comprise a fold-yard for the beasts which was surrounded by byres, haybarns, a residence for the granger and any other monastic staff, and bothies for the herdsmen. Also present were bercaries or sheep farms, supporting flocks of 200–300 sheep. The granges engaged in a spectrum of activities, and Raistrick described how they specialised according to the resources of the locality: 'At an early date Fountains Abbey had granges which specialised: Brimham in iron making and lead smelting, Bradley in iron making, Kilnsey in sheep and wool production, Stainburn in arable crops, and Horton in horse rearing.' The grange system of the Cistercians was so successful that it was adopted by some other orders, like the Gilbertines.

The great period of monastic expansion ran from the conquest to the second quarter of the thirteenth century, then settled on a plateau of prosperity, but a long slope of decline spanned the years from the mid- fourteenth century to the Dissolution. In the case of the Cistercians, their spectacular early appeal to benefactors centred on their austerity, their readiness to colonise the most barren of places, and their renunciation of wealth. According to Raistrick, the compromising of these principles and the breaking of some of their own rules derived partly from their enormous success, particularly in the field of wool production, from the desire to expand, enlarge and beautify their abbey buildings, and from the temptations posed by the presence of wool buyers at their gates. The reasons for the decay of monasticism in general were various. Although Scottish raiding could be troublesome and occasionally disastrous in the North, it could not generally retard a movement that was buoyant and ambitious. During the phase of monastic growth, there had been no shortage of landowners in Yorkshire who were ready and able to provide aspiring communities of monks with generous endowments of land. Sometimes cruel, occasionally saintly, these benefactors were invariably devout; many had good cause to fear for their mortal souls and some had younger sons or brothers who could be well placed as abbots- or priors-to-be. Gradually, however, the stream of endowments dwindled, parish churches became more attractive as targets for patronage, and the reputation of some monastic houses was tarnished by rumour, fact and bad example. The public image of the monasteries was not helped by the attractive example of the Grey and Black Friars, who became an important influence in thirteenth-century England, winning great popular support and royal patronage and establishing four houses in York. They based themselves in towns, but did not own their buildings, which were held in trust for them by corporations of citizens. Not surprisingly, the zeal and poverty of the friars reflected on the affluence, remoteness and possible corruption of many monastic communities, and some monks, like the Cistercians at Scarborough, sometimes strove to keep them away.

Economic and environmental problems also eroded the monastic fortunes; after its arrival in 1348, the Black Death could strike blows from which a foundation would never recover. In 1349 Meaux Abbey lost 33 members of its 50-strong community; Fountains, which had supported a community of at least 250 in its heyday, had a complement of only 44 by 1380; and the comparable figures at Rievaulx were about 700 and 18 – a cataclysmic collapse. Since the standards of cleanliness in the monasteries were far higher than those prevailing outside, it is perhaps surprising that they were hit so hard by the pestilence. Even so, with their magnificent buildings and great estates, the monasteries could have recovered – like most towns and villages did. Recruits were few, however; the monastic vocation had lost its appeal and, after the plague years, attractive agricultural tenancies were available to all able-bodied survivors. There were also severe problems on the economic front. The Cistercians were the

largest wool producers throughout the thirteenth century, but taxation, the failure to exploit opportunities in spinning and weaving, and competition from other lay and monastic producers all helped to erode the profits, and, at the end of the thirteenth century, Fountains was heavily in debt to Italian merchants who had bought futures in wool. There was also a ever-growing tendency for grange land to be let out to lay tenants – people who found themselves in an advantageous economic position as a result of the labour shortages caused by the Black Death. In Swaledale, this process began at Oxnop in 1301; in Wharfedale, Bolton Priory farmed out its manors in the fourteenth century and was virtually insolvent in the fifteenth. One method of raising capital in hard times was the sale of 'corrodies', accepting a grant of cash or land from an individual or a family in return for the promise of accommodation and succour for the corrodians in their old age. Of course, the system would backfire if too many corrodians lived too long, and in 1321 Kirkham Priory had burdened itself with 22 corrodians, almost equalling the number of canons. This venture was the result of cash shortages caused by an ambitious building programme. Kirkham owed £1,000; by the end of the thirteenth century, Kirkstall had run up debts of £5,000 and, in 1274, Fountains was £6,373 in debt as a result of overambitious building works.

At the time of the Dissolution, the monasteries of Yorkshire retained their magnificent buildings and most of their estates, but many of their communities were severely depleted. Corridors and dormitories which had once been packed with monks or lay brothers were now mere shells, echoing to the occasional footstep, while the naves of the great abbey and priory churches remained largely empty during mass. Although Bolton had doggedly preserved a community of about 15 canons, Rievaulx had only an abbot and 21 monks in 1538; Guisborough, once so rich, had a prior and 24 canons in 1539; Fountains had an abbot and 31 monks. Kirkstall, which had supported 36 monks and many more lay brothers, had only 17 monks and 6 lay brothers after the assaults of the pestilence, but had partially recovered to accommodate 31 monks by the Dissolution. For all these problems, Prior Moone launched the building of a west tower at Bolton in about 1520. The project, probably in imitation of the spectacular Perpendicular tower built at Fountains a few years earlier, was incomplete when the Dissolution swept the community away.

Despite the periodic eruption of conflicts between the peasantry and the monastic authorities, and, eventually, the emaciated condition of the monastic movement, the Dissolution was often marked by sorrow and resentment in Yorkshire, if welcomed in the South. A rebellion by 'pilgrims' in Lincolnshire spread across the Humber, and in 1536 pilgrims from various parts of the North rallied near Market Weighton and were led towards York by the lawyer Robert Aske.

Here they were welcomed and marched south to Pontefract, gathering support from nobles, churchmen and soldiers. As well as supporting the Roman Catholic cause, the marchers on this, the 'Pilgrimage of Grace', were fearful that southerners would acquire the monastic estates, milk the North of its wealth, and so leave it weak, sandwiched between Scottish raiders and southern guile. There were also fears that, with the loss of the priories, many communities would be deprived of their churches and priests – an understandable concern in the remoter areas. Such places were also experiencing great economic hardship and were particularly sensitive to anticipated changes in taxation or landlords. However,

The Augustinian priory at Bolton in Wharfedale was founded in 1154–5 and experienced as chequered history, punctuated by Scottish raids, plague and debt

the pilgrims foolishly negotiated with the Duke of Norfolk at Doncaster and despatched representatives to speak to the king at Windsor. Gradually, their host of 30,000–60,000 men melted away, then promises and pardons were forgotten as the king's army invaded. The five northern counties were placed under the jurisdiction of the Council of the North, created to maintain royal power and surviving for a century.

Today the monasteries are seen as jagged, empty shells. Fragments of ruined masonry stand upright like rotting teeth. These fragments bear architecture and decoration in the styles of the four centuries during which monasteries were almost supreme in many parts of Yorkshire. It is interesting to see how the different personalities and outlooks of the various orders are encapsulated in the ruins that they left behind. The great Cistercian colonisation of Yorkshire overlapped with the Norman and Early English building transition. The order was swift to adopt the graceful, uncluttered austerity of the first Gothic style, and the change-over from the Romanesque to the Gothic can be seen at several Yorkshire monasteries, like Fountains and Rievaulx. It is particularly apparent at Kirkstall, where the heavy, round-headed Norman arches surround the cloister and define the windows above, but where the constructional arches are pointed and slender shafts cluster round the piers of the nave.

Stately Early English architecture proclaimed the grandeur of the monastery at Rievaulx, in its day the most celebrated of the Cistercian houses in Yorkshire

The Cistercian houses supported a large contingent of lay brothers, the labourers and stockmen of the community, who were kept illiterate, worshipped in the morning and evening, but were otherwise engaged in farming and manual tasks. The Cistercians devoted the western range of their monastic buildings to the accommodation of lay brothers and housed their abbot elsewhere, whereas other orders allocated the western range to the abbot. As the number of lay brothers dwindled in the Cistercian foundations, their refectory, common room and dormitory space was often converted into offices or guest accommodation. The most conventional Cistercian layout is seen at Roche, traced out on the turf by the wall footings, with the rectangular cloister bounded to the north by the cruciform church, to the east by the chapter house, with the warming house, frater, or refectory, and kitchen to the south, and the lay brothers' frater, with its dormitory above, to the west. The abbot's lodging and kitchen were set further from the cloister, to the south. Here a stream running beneath the buildings of the southern claustral range provided a convenient drain, but it is at Fountains that the most skilful exploitation of a river is found, supplying cool water and then flushing the drains as it flowed along. The cramped nature of the hillfoot site at Rievaulx demanded a modification of the conventional plan, so that the church lies to the east of the cloister and is almost orientated to north and south rather than to east and west.

The most distinctive of monastic layouts can be seen at Mount Grace Priory. Here the architecture was gaunt in comparison to the restrained but graceful work of the Cistercians and the quite lavish decoration of some Benedictine and Augustinian houses, reflecting the austerity of the Carthusian order. The modest, ruined church is in the Perpendicular style, the priory not being founded until 1398. The northern half of the site is occupied by the diamond-shaped cloister, which was surrounded on three sides by cells set in tiny gardens. The Carthusian monks spent most of their time in isolation in their cells, reading, praying and then working in their plots. One can still see the little hatches beside the cell doors, through which meals were passed to each lonely monk.

While the ruined monastic buildings are the most obvious survivors of monasticism in Yorkshire, whole countrysides have been reshaped by the monasteries. The Cistercians were the most influential order in Yorkshire and, at first, their influence was partly destructive. As they erupted into countrysides that were still scarred by the Harrying, they encountered peasants struggling to reclaim farmland from the waste. The Cistercian craving for isolation had severe consequences for dozens of innocent rural communities, whose inhabitants were evicted, often with little concern for their future. The clearances tended to come in two phases: the first, and least destructive, involved the removal of settlements which were considered to lie too close to the abbey building sites; while the second involved the depopulation of outlying villages, hamlets and farms during the expansion of the abbey granges. Early victims of Fountains were nearby Cayton and Herleshow, and the Abbey of Meaux was responsible for the extinction of its namesake, the settlement of Melse being depopulated and re-emerging as the abbey's North Grange. In the case of Byland village, the transition was gentler; during the period when the monks were living uncomfortably close to Rievaulx, Byland village was removed, but a new, purpose-built settlement of Old Byland was created for the evicted community. Of course, the monks were not Cistercians at this time, but belonged to the order of Savigny. It is possible that the abbeys of Jervaulx, Kirkstall, Meaux, Rievaulx, Byland and Fountains were together responsible for the eventual removal of more than 60 peasant settlements. At Kirkstall, shortly after the foundation of the abbey, the monks were disturbed at their worship by sounds from a service at the nearby parish church of Barnoldswick. They destroyed the church and raised its dependent chapels of Bracewell and Marton to the status of parish churches. The Cistercians did not entirely monopolise the injustices: when the canons of Kirkham were granted one estate, it is recorded that they evicted the peasants, seized their produce and stored it in a barn. The peasants burned the barn and killed three lay brothers before being arrested.

Having obtained their great domains, the Cistercians set about exploiting them in an advanced and business-like manner. Most of the endowments of land were given with the intention that they would provide food and a little income for the devout communities, but the monks developed them to their full commercial potential, which often involved the removal of subtenants and direct management. In many places, the production of wool, which had an ever-buoyant market and was light in its demands for labour, was the chosen option. The Yorkshire flocks yielded wool which, though rather coarse, was very strong. With their gigantic estates, the Cistercians were able to avoid the localised problems of the build-up of disease and mineral deficiencies in a pasture, for the flocks could be kept on the move. This also allowed for seasonal migrations of the flocks; for example, Fountains could winter its flocks in sheltered Nidderdale and move them to the high grazing above Wharfedale in the summer. The monks also explored the treatment of some ailments and were expert in quality control and marketing techniques, sometimes buying up fleeces from lay producers for resale. Other denominations, like the Austin canons of Bridlington and Bolton, or the Gilbertine canons of Malton, were also involved in the wool trade, but wool did not monopolise the monastic farming. Estates were worked from granges manned by lay brothers and, beside these farms, there could be plough-land in compact fields which lacked the strip patterns of peasant agriculture. In some places, like Swaledale, vaccaries were common, while Bardney Abbey had a horse farm in this dale at Healaugh. Fountains established some granges as mining centres, as at Ainley in Elland, and Rievaulx had similar establishments in the Bingley area. Although the monks were always keen to 'round off' their estates by gaining control of enclaves and salients, the pattern of the holdings was largely determined by the accidents of sale or endowment. These would reflect which lords favoured which order or foundation, who felt the need for prayers to be said for his (or her) soul, who needed to raise some ready cash to finance a journey to the crusades or to pay their taxes, and so on. As a result, the property mosaic could be very complicated. In Nidderdale, there were areas of royal forest as well as lay manors and estates

belonging to both Byland and Fountains. In Swaledale, there were lay manors and also the properties of the nunneries of Marrick and Ellerton, the abbeys of Coverham, Bardney, Rievaulx, Jervaulx, Easby, and the priory of Bridlington. There were many places where monk and peasant, lay brother and knight shared roads and boundaries, and fierce disputes were not uncommon. Many of the most heated conflicts concerned the impounding of stock in abbey pounds. In 1241, for example, Rievaulx gained pasture rights in Upper Swaledale and impounded any cattle found on the traditional grazings. Two local men broke into the pound to retrieve their cattle, stole some dogs and kidnapped a keeper. Similarly, in 1252 two men from Healaugh kidnapped the Abbot of Rievaulx's granger at Keld and broke open the pound to release their stock. It was alleged in the following century that members of the Fountains community were liable to attacks by armed bands in Craven, while there were various cases of monks being accused of driving sheep from common grazings, breaking mill-dams or using sharp practices or false evidence to prise out the occupiers of desirable holdings. Occasionally, the monks were in conflict with each other, like the incident in Littondale in 1279, when the monks and lay brothers of Salley destroyed a mill belonging to Fountains because of a dispute over water rights. The lack of mapping skills, the vague nature of written boundary descriptions, complicated customary rights, and the tendency for monks to forge charters ensured a lively history of boundary disputes.

Monastic management and independence in the countryside waned before it disappeared, but the centuries of monastic control left an almost indelible mark on some landscapes. Parts of the boundary wall of Fountains Abbey can still be seen amongst the less distinguished neighbouring field walls. Monastic granges frequently survived as lay farmsteads and sometimes grew into villages or hamlets, like the ones mentioned in Upper Nidderdale, or the hamlet of Kildwick Grange, near Skipton, which grew from a grange of Bolton Priory. In the Vale of York, granges seem often to have been established on the glebe lands of expropriated churches and then to have expanded their operations into the surrounding areas.

'Grange' place-names can be confusing. Sometimes the *grangia* was a monastic farm, but the word was also used to describe a simple barn or to dignify eighteenth- and nineteenth-century houses with no monastic connections. Portions of the architecture of granges can be seen at places like Kilnsey and Ramsgill, and the remains of lodges are still being discovered. In some places, countrysides or farmsteads result from the removal of larger settlements that were establishing themselves when the Cistercians moved in; other gaps in the village maps denote later lay clearances for sheep pasture, once the monks had shown just how profitable the wool trade could be.

REFERENCES

L. G. D. Baker, 'The foundation of Fountains Abbey', *Northern History*, 4 (1969), pp. 29–43.

L. G. D. Baker, 'The desert in the North', *Northern History*, 5 (1970), pp. 1–11.

T. A. M. Bishop, 'Monastic granges in Yorkshire', *English Historical Review*, 51 (1936), pp. 193–214.

J. Burton, 'Monasteries and parish churches in eleventh- and twelfth-century Yorkshire', *Northern History*, 23 (1987), pp. 39–50.

J. Burton, *The Religious Orders in the East Riding of Yorkshire in the Twelfth Century*, East Yorkshire Local History Series 42 (Beverley: 1989).

M. Chibnall, 'Monks and pastoral work: a problem in Anglo-Norman history', *Journal of Ecclesiastical History*, 18 (1967), pp. 165–72.

B. Dobson, 'Mendicant ideal and practice in late medieval York', in P. V. Addyman and V. E. Black (eds), *Archaeological Papers from York Presented to M. W. Barley* (York: 1984), pp. 109–22.

R. A. Donkin, 'Settlement and depopulation on Cistercian estates during the twelfth and thirteenth centuries, especially in Yorkshire', *Bulletin of the Institute of Historical Research*, 33 (1960), pp. 141–65.

R. A. Donkin, 'Cattle on the estates of medieval Cistercian monasteries in England and Wales', *Economic History Review*, 2nd series, 15 (1962) pp. 31–53.

R. A. Donkin, 'The Cistercian grange in England in the twelfth and thirteenth centuries, with special reference to Yorkshire', *Studia Monastica*, 6 (1964), pp. 95–144.

R. A. Donkin, 'The English Cistercians and assarting c. 1128–c. 1350', *Analecta Sacri Ordinis Cisterciensis*, 20 (1964), pp. 49–75.

R. A. Donkin, 'The Cistercian order and the settlement of Northern England', *Geographical Review*, 59 (1969), pp. 403–16.

R. A. Donkin, *The Cistercians: Studies in the Geography of Medieval England and Wales* (Toronto: 1978).

C. V. Graves, 'The economic activities of the Cistercians in medieval England (1128–1307)', *Analecta Sacri Ordinis Cisterciensis*, 13 (1957), pp. 3–60.

R. Hoyle, 'Monastic leasing before the dissolution: the evidence of Bolton Priory and Fountains Abbey', *Yorkshire Archaeological Journal*, 61 (1989), pp. 111–37.

M. D. Knowles, *The Monastic Order in England, 940–1216*, 2nd edn (Cambridge: 1963).

J. McDonnell, 'Upland Pennine hamlets', *Northern History*, 26 (1990), pp.20–39.

A. Raistrick, *Monks and Shepherds in the Yorkshire Dales* (Yorkshire Dales National Park Committee: 1976).

A. Raistrick, *Malham and Malham Moor* (Clapham: 1983).

R. B. Smith, *Land and Politics in the England of Henry VIII* (London: 1970).

B. Waites, 'The monastic settlement of North-East Yorkshire', *Yorkshire Archaeological Journal*, 40 (1959–62), pp. 478–95.

CHAPTER TWELVE

The Medieval Countrysides

YORKSHIRE DID NOT HAVE one medieval landscape, but several. If one regards the medieval period as running from the conquest to the Dissolution, then a time-span of almost half a millennium is involved – and, in the course of these five centuries, all countrysides were evolving. In some places, the evolution was gradual, so that Norman or earlier patterns and practices were still recognisable when the medieval period drew to its close. In other places, however, changing values and practices had completely transformed the Norman countrysides by Tudor times. The story of the countrysides involved an interaction between the forces of colonisation and decay – and, then as now, decay could be the consequence of ill-considered colonisation. In general, Yorkshire, like other regions in England, experienced a phase of sustained expansion in the years up to about 1300, followed by crisis, disaster and change.

The Domesday countrysides of Yorkshire were patchy and variable, with working lands juxtaposed with uncolonised waste and ravaged holdings, and with lay lords and monks engaging in reclamation and commercial farming enterprises of different kinds, although the detail of the operation of field systems is still debated. Across the face of Yorkshire, local conditions of culture and power were expressed in the visible landscape. The countryside at Grassington, where surviving medieval aspects were recorded in a survey in 1603, can serve as an example. Between the lower end of the present village and the Wharfe, lay Sedbur Field, one of the community's three open fields; to the east and west of Sedbur Field lay two larger open fields, East Field and West Field. In any season, one of the open fields would have been rested and grazed as fallow, and crops of oats, barley and legumes would have been grown

in the other two fields. The open fields were walled and divided into more than 800 'furshotts', strips or selions. Between the southern margins of the fields and the river ran a ribbon of meadowland which produced hay for winter fodder; like most other haymeadows, this would probably have been divided into tenant strips or 'doles', just like the arable fields. On a hillside in the south-western corner of the parish lay Grass Wood, an ash-dominated wood measuring about a mile by half a mile (1.6 by 0.8km); it still exists today, serving as an unusually important nature reserve. Since the wood is growing over Iron Age dwellings and a small hill-fort, one suspects that the area was overrun by woodland during the Dark Ages; in medieval times, the wood was an exclusive reserve for the lord's deer. Most of the land on the limestone slopes above Grassington that had been divided into small fields in Iron Age and Roman times existed only as common pasture and rough upland grazing in the medieval period, with the common pasture and high grazings of the Out Moor comprising about three-quarters of the township lands. Between the common pasture and the open fields were a number of old intakes, enclosed from the waste with the lord's permission. One of them yielded coppice timber, which was useful for making temporary fences and for construction work and which could not be taken from Grass Wood. The moor yielded peat for fuel and many other things that were vital to the peasant economy.

It would be very hard to discover the details of the evolution of farming patterns in Grassington during the Middle Ages, but in Yorkshire generally the retreat of agriculture which was caused by the Harrying was followed by a period of high population growth and agricultural expansion.

Crosses were ubiquitous symbols of the medieval countryside, though it is not always easy to deduce their significance. They are remarkably numerous on the North York Moors, where some at least may have served as waymarks. Young Ralph Cross is probably medieval and was recorded in 1200. It was repaired following damage by a climber who was pursuing the tradition of looking for alms on the top of the cross

The limits of Saxon farming were reached and then surpassed, creating pressure to take in new areas of ploughland and pasture. Assarting, the clearing of woodland, heath and moor from the waste, often provided an escape valve for population pressures and led to the unveiling of countryside which had not been tilled or grazed as

pasture since Roman times. Medieval assarting activities are frequently revealed by place-names. In Yorkshire, the most widely used name was 'rod', 'rode' or 'royd', but names like 'rodeland', 'ridding', 'breck' or 'broke', and 'stocking', 'stubbing' or 'stocks' were also used. Assarting could take several different forms. It could be a collective endeavour by the peasants of a township, involving the creation of communal ploughland which might add a third open field to a two-field system or enlarge existing open fields. It could involve freeholders, who bought licences to assart sections of the waste from a manorial lord, producing enclosures which often survive as irregular hedged or walled fields today. In a unique charter of 1259, Henry II charged 4*d.* for every acre cleared for assarting in the Drax area in the south of Yorkshire, and this large-scale clearance probably resulted in considerable immigration. Assarting was also accomplished by manorial lords and substantial freeholders. Certain families specialised in assarting and made good livings from the process; they included the Arkel family in Nidderdale and the de Assartis or del Sarte family in the Batley and Pudsey localities. While assarting was the most obvious means of obtaining the extra land to support a rapidly growing population, it could also be a contentious issue. This was because the land concerned was seldom worthless, for the medieval waste was not waste in the modern sense of the word and often existed as common wood pasture, heath grazing or private forest. Assarted land had to be enclosed, normally by hedges, which could be a problem when the assart covered old common land with attached grazing rights – which were often considered to prevail once harvesting was over. For example, at Killinghall, in Nidderdale, one Henry, son of Walter, had agreed with the community that he would fence his assarted land while corn was growing, but open it to common grazing after the harvest. In 1344 he was accused of keeping the land enclosed for his own sheep and cattle. Similarly, two years later, William of Baildon complained that local people had broken down his hedges at Oxenhope and allowed their cattle to devour his corn and meadow-grass.

The medieval assarter is sometimes regarded as a resourceful individualist, comparable to the

The Barden Tower was one of six lodges built for foresters in Barden forest. The house was rebuilt in the fifteenth and seventeenth centuries and repaired in 1774

log-cabin-building pioneer of North America, but detailed work by Harrison showed that: 'The one kind of settlement for which no evidence has so far come to light in the North York Moors is the piecemeal assarting by individual peasant households which is so often held up as the norm.' Instead, he found that a variety of classes were involved in assarting, with freeholders playing a significant role:

> while seigneurial planning was an important factor in some areas, notably in royal and private forests, the great lords who dominated the area also allowed important freeholds to develop within their territories partly through unrestricted assarting. A clientele of substantial freeholders was vital for the functioning of a great honour and for the humble jobs in local administration generally.

Rather than assarting being a peasant-led assault on primeval woodland, the process was often organised by leading figures in society; the Abbot of St Mary's, York, organised assarting in Farndale, on the North York Moors, while the king's bailiffs assarted Goathland in the same region.

While many of Yorkshire's isolated farmsteads are a product of parliamentary enclosure in the eighteenth and nineteenth centuries and some have evolved from granges, there are also some which originated in medieval assarts. The farm at Lower Hazel Hirst, in Northowram township on

the vast manor of Wakefield, has been given as an example. In 1311, a licence to assart 11 acres of land here was given to one Richard, son of Jordan, who then sold the cleared land; five years later, farmland with buildings was recorded here. Since most assarts lie on the far margins of village lands, it would often have been convenient for the holders of such clearings to move out and settle upon them. Assarting is generally associated with an expansion of the cultivated and settled area from the lowlands into the uplands. There were also cases where assarting involved a movement down the slopes. Jennings described how, in Calderdale, in the southern Pennines, people lived in villages and hamlets at heights of 215–75 metres; after 1250, they began to clear the woodland on the slopes below, so that, by 1300, they were wrestling with the problem of draining the flood-prone valley bottoms. He thought that 'royd' names seemed to relate specifically to the clearance of woodland, with many of these names in Yorkshire dating to the period 1275–1325. Holmfirth was a colony of the older, higher settlement of Holme, established in what had been Holme's common wood.

In general, assarting benefited the community

and its lord. The latter was able to increase his rent revenue, profit from the sale of licences, and fine tenants who had failed to hedge their holdings. Not all the enclosures were licensed and, periodically, there would be crack-downs on unlicensed felling or burning. As the population grew, so too did the urge to assart; there seems to have been a rapid increase in pressure upon the waste in various parts of Yorkshire at the start of the fourteenth century. The supply of lightly used countryside was beginning to dry up. After the Harrying, there had been tracts of empty land available for hunting forests, but the shortage of economic resources began to press upon the expanses of aristocratic recreational land.

The term 'forest' was used in a legal rather than a descriptive sense: the royal Forests included much unwooded land and some that was under peasant agriculture. Inside the Forests, however, the forest laws, which penalised the unlicensed extension of farming, prevented the protection of crops against the game, imposed harsh penalties for poaching, and demanded the maiming of peasant dogs, prevailed. Norman Yorkshire contained extensive Forests. On and around the North York Moors, there were the Forests of Scalby, Whitby and Pickering, the latter some 16 miles (25.7km) long and several miles wide. Near York was the forest of Galtres, and further west, in Nidderdale, the great Forest of Knaresborough. There were also the chases hunted by the nobility, like Arkengarthdale and New Forest in Swaledale, which were hunted by the lords of Richmond, while the earthen banks of the pale of Robert de Brus's hunting estate at Guisborough can still be traced. In 1227, as previously noted, one Ranulph, son of Robert, claimed that his ancestors had created the village of Bainbridge, in the Forest of Wensleydale, to accommodate 12 foresters, each with a house and 9 acres of land. Raistrick thought that Buckden and Healaugh had also originated as forest lodges. Within the Forests, the small peasant communities farmed subject to the restrictions associated with Forest life, often under the authority of a warden, his foresters and woodwards, and the local court or wood-mote. Each community was likely to be burdened with special responsibilities, sometimes the care of hunting dogs and often the building and maintenance of woodland boundaries. Raistrick wrote that:

> As the forests included the villages and hamlets within their area, often with a numerous total population, the common rights that these people would have enjoyed in a free village were usually secured to them within the forest. Other rights were added theoretically as a recompense for damage done by deer and other game, with the result that within the forest there was constant conflict of royal prerogative and the people's common rights.

Whether or not a community lived within or far from a Forest or chase, it depended heavily on woodland resources. Timber was needed for house-building and fencing, light coppice timber for poles, wattle and posts, brushwood for roofing and fuel, while pigs had 'pannage' on the woodland floor. Some woods were the exclusive hunting preserves of the nobility, though rights of pannage were often allowed at certain seasons. In some townships, the lord had his own piggery in the demesne wood, though peasant life would have been impossible without the existence of a common wood or the upholding of commoners' rights in another wood. Place-names like Swincliffe (Nidderdale) and Swinstie (various) probably reveal the locations of manorial piggeries. Holly was an important source of winter fodder and could be grown in the secondary layer of a wood or in special holly woods, sometimes revealed by the name 'Hollins'. Being so valuable, woods merited careful management and protection. They would be ringed by banks, ditches and palings; within the wood, young coppices needed protection from browsing, though grazing was possible amongst stands of pollarded trees. In winter, the pressures of use on fallow arable land and meadow were great, and the sheltered winter grazings provided by woodland played a vital part in the peasant economy.

Woodland was a vital asset, but, as colonisation proceeded, it tended to become increasingly confined to the steeper slopes, heavier clays or hungrier sands – so that the new farmland created by assarting was less and less likely to comprise soils of sufficient quality for use as semi-permanent ploughland. Nevertheless, the pressure to create more farmland in the twelfth and thirteenth centuries led to the removal of a high proportion of Yorkshire's woodland; even had they wished to stem the process, there is probably little that the lords could have done to arrest the tide. One year, a peasant might graze some beasts at the edge of a wood and the animals would nip off the tips of seedlings. A little later, he might surreptitiously fell a few trees and, before very long, ploughed fields would appear. One often reads of people being fined for illegal fellings, but the lord would usually have recognised or supported the inevitability of the activity; what he was really doing was taxing and renting, rather than fining as the term is understood today. In the Forest of Galtres, the men of Easingwold and Huby had common for oxen, cows, horses and mares, and also swine, except for a fortnight on either side of Midsummer Day, when pigs were excluded to protect the newly born fawns. Sheep and goats were banned. The men had no rights to take timber for house-building or hedge-making. McDonnell points out that Domesday Book does not reveal this Forest as being heavily wooded; in 1251 a justice of the Forest was ordered to kill and salt 100 bucks from the forest, but, by the second half of the thirteenth century, the woodland cover may have become too fragmented by clearances to allow Galtres to function as a hunting Forest – though the role was not formally surrendered until the reign of Charles I. The existence of Forests as exclusive hunting reserves for the monarch did not last beyond the reigns of the Norman and Plantagenet kings; John hunted in Knaresborough Forest on several occasions, but the pressures on the Forests soon became irresistible as land was cleared, domestic livestock invaded, and cattle farms and horse studs became established. By the fourteenth century, deer hunting had become a stylised ritual and the deer were being kept in the closer confines of a park.

Lords of substance had estates scattered across the length and breadth of the kingdom and would proceed from one to another accompanied by a retinue of companions, officials and servants. When the procession arrived at a castle or manor, ready supplies of hunting and fresh meat were expected. The pressures on the game resources of the Forests and chases made the creation of deer parks extremely popular in thirteenth-century England; more than 70 examples were created in the West Riding of Yorkshire alone, and there were 67 in the North Riding. The deer park was a relatively compact area which was surrounded by banks and ditches punctuated by deer leaps, which allowed the deer to enter but not to escape. In the larger parks, the animals may have been hunted inside the park; in the smaller ones, they were probably released and pursued across the surrounding farmland. If the lord had demesne woodland available to house his new park, then the creation of a deer park was a minor worry for the neighbouring peasantry, but occasionally he 'emparked' peasant land. Around 1173, William de Stuteville earned the resentment of the local community when he emparked common grazings to form Haverah Park, near Knaresborough. The park covered some 2,250 acres and was unusually large. At Fewston, near Harrogate, are the remains of 'John of Gaunt's castle', in fact, the ruins of a hunting lodge. Two other deer parks came into being rather later, at nearby Bilton and Haya, which were stocked by deer from royal reserves at Pontefract and the Forest of Galtres. Such long-distance exchanges of deer were not uncommon in the thirteenth century, again suggesting that the animals were less numerous than before – and probably much less common than today, when their population is expanding.

The most eye-catching survival from the days of deer hunting is the Barden Tower in Wharfedale, a medieval hunting lodge extensively repaired in 1774. The Forest of Barden was granted to Robert de Romille in 1066 by William I, and to Robert Clifford in 1310 by Edward II. Six lodges were established on the Clifford estates, with Barden being developed in the late fifteenth and sixteenth centuries as a major manorial centre. All the lodges were on the valley sides, occupying good vantage points. They housed keepers and officers of the Forest who were responsible for protecting the various resources of the Forest – game, timber, bark for tanning, pannage for pigs, beeswax and honey. Beaumont notes how the lodge sites progressively became farming settlements as woodland pasture was leased for grazing

THE YORKSHIRE COUNTRYSIDE

and cleared for cultivation. The accounts of Skipton Castle show that, in 1322–3, the lodge sites were yielding income as vaccaries, though the lords of Skipton retained control at Barden, where the lodge was rebuilt in stone in 1495 as a prestigious residence and seat of the manor court. Associated with it were fish-ponds, a rabbit warren and a deer park protected by a pale. In about 1650 the Barden Broad Park was enclosed by a wall almost six miles (4km) long, and the preservation of deer had priority there until the end of the seventeenth century. The lodges at Drebley and Gamsworth became farming settlements; the loose hamlet with its own open-field system contracted to a single farmstead at Gamsworth, while there were seven households in Drebley hamlet in the eighteenth century.

Several deer parks had disappeared before the end of the Middle Ages. Some lingered and re-emerged as landscape parks; the landscapes created by Repton and his followers in the late eighteenth and early nineteenth centuries were really re-creations of the lawns and tree patterns of the medieval deer park. As early as 1307, a herd of 70 mares had displaced the deer in the park at Skipton Castle. Much of the boundary wall that enclosed the deer park of Ravensworth Castle, near Richmond, still survives. Williamson notes that: 'Parks were expensive both to create and maintain. In particular, the perimeter pale had to be constantly repaired, not only to keep deer in but also to keep poachers out. The threat here came not so much from starving peasants as from political rivals, for "park-breaking", hunting openly in an enemy's park, was the supreme affront.' He adds:

In the later Middle Ages the number of parks in England declined. As the economy slumped and real wages rose in the decades following the Black Death, parks simply became too expensive for most members of the gentry. The climatic deterioration which accompanied these economic changes may also have played a part. Fallow deer, a Mediterranean species, have little subcutaneous fat and an inadequate coat to withstand what became a series of unusually bitter British winters. The epidemics that decimated herds of cattle and flocks of

sheep in the fourteenth century may well have taken their toll on deer, too.

The expansion of peasant farming in Yorkshire took place in conditions of increasing difficulty. We can identify three great crises in the history of farming in Yorkshire – there were probably prehistoric crises, too, but they are harder to identify. The first crisis resulted from over-exploitation in Roman times, as noted above. The second occurred between the late thirteenth century and 1348. The third crisis is occurring today, with all the established portents of soil destruction and environmental damage. As the population grew in the twelfth century, communities colonised neglected lands, but the resources of the waste were gradually exploited to their limits. On the better lands, arable farming could support more people than could livestock farming – but, as ploughland expanded at the expense of pasture and wood, less manure could be produced and fertility declined. With yields falling and populations rising, desperate peasants then tried to bring marginal lands under the plough and overstocked their pastures – and the weak lands became weaker. Meanwhile, inflation was aggravating the situation and a marked worsening of the climate was becoming apparent. Relics of the frantic urge to produce more food are still visible in the countrysides of the Dales. Seen in winter with the grass closely cropped, dusted with snow and the slopes lit by the low, slanting rays of the sun, it becomes apparent that huge expanses of the valley slopes are patterned with the faint corrugations of ridge and furrow. These are not the bold patterns to be seen on the lowland areas of former open-field land, but relics of the shortages which impelled peasants to till ground which would never normally have been ploughed. Far more dramatic are the 'strip lynchets' or 'raines', artificial cultivation terraces which pleat the hillsides in upland limestone country, as near Linton and Burnsall, in Wharfedale. They are entirely artificial, created by unidirectional ploughing, which turned all the sods downslope, with the soil being retained by rows of gathered boulders which formed the risers in the hillslope staircase. In this thinly soiled and steeply sloping country, the creation and working of raines would have been countenanced only in times of severe hardship, which makes it seem likely that the horizontal flights of raines are products of the thirteenth and early fourteenth centuries. Other good examples are displayed at Carperby, in Wensleydale, near Kirkby Stephen, and in the Malham area.

Scottish raiding was a factor in the problems of fourteenth-century Yorkshire, but the greater troubles that afflicted Yorkshire were also afflicting other English communities which lived far from any threat of Scottish incursions. Yorkshire was caught in the vice of environmental decay, some of the trouble stemming from an overworking and exhaustion of countryside resources, others from the inevitable consequences of climatic change. The northern farming environments were severely hit by these changes. Working on the Lammermuirs in south-east Scotland, Parry found that, as the climate worsened, the average frequency of crop failure rose from less than one year in twenty to one in five by the mid-fourteenth century and to more than one in three in the mid-fifteenth century. He wrote: 'During the optimum of 1150–1250 the climatic limit to cultivation in the Lammermuirs stood at about 450 m OD. By 1300 it had fallen to perhaps 400 m OD, although all but the summits of the hills may still have been suited to cereal cropping. Thereafter, the late medieval deterioration promoted a further fall of 75–90 m and the uncultivable upland core was more than doubled in area.' He concluded: 'It is evident that early cereal cultivation on marginal, maritime uplands was particularly sensitive to changes in summer warmth, summer wetness and exposure. When secular trends in temperature and rainfall are resolved into accumulated summer warmth, frequency of crop failure and potential water surplus, it is evident that the theoretical limits to cropping of oats may have fallen about 140 m in south-east Scotland between 1300 and 1600.' There is evidence of a comparable deterioration for Yorkshire. Shaded by the ever-rolling clouds and lashed by rain, crops of wheat, barley, oats and rye rotted in the fields. Meanwhile, diseases of sheep and cattle erupted in the sodden moors and pastures. Lambs perished in spring blizzards, and there was a shortage of hay to sustain their weakened mothers. Fieldhouse and Jennings have shown how the sheep flock at Bolton Priory was reduced in one year

Flights of strip lynchets give this hill near Linton, in Wharfedale, its furrowed brow

Traces of conventional ridge and furrow ploughing near Bolton Priory. Note the 'reversed-S' shape of the ridges, caused by the need to begin to turn the long plough team as the headland was approached. During a thaw, the patterns of medieval ploughing can be seen throughout the dales

Traces of medieval ploughing are clearly seen beyond the farmsteads on the outskirts of Kettlewell. While normal plough ridges have a dome-shaped cross-section, these have a wedge-like profile. This phenomenon is still to be explained, but could result from the plough being worked only in a downhill direction on these steeper slopes, producing scarps rather than ridges

from 3,000 to 1,000 animals, while the cattle herd was reduced from 225 beasts in 1318–9 to only 31 in 1321. At the same time, the *Lanercost Chronicle* described how a plague amongst the oxen in the North obliged farmers to harness horses to their ploughs. A poor harvest in 1314 was followed by a dreadful summer the following year, when crops rotted in the waterlogged fields. Starvation and disease were the normal companions of a bad harvest, and the price of bread-grain increased eightfold in some districts. Even oats, normally well adjusted to the conditions of northern farming, were being sold at more than three times the normal price. The Scots fell upon many parts of Yorkshire, exploiting their victory at Bannockburn in 1314, to increase the misery. In the first half of the fourteenth century, the assarted area increased by a quarter in some places, but the waste was no longer able to underwrite the demands of a swollen population.

Fate found a remedy for all the the problems associated with excess population, but it would be hard to imagine one more cruel. I have already mentioned how the Black Death affected the monasteries. Arriving in England in 1348, it spread rapidly to all corners of the realm; it had two principal forms, was carried by rat fleas, and was usually both painful and fatal. Traditionally, it has been argued that the Black Death exterminated between a third and a half of the English population by the end of the fourteenth century, and the Yorkshire evidence would seem to support this high level of mortality. Periodically, and unpredictably, the pestilence would return, remaining a threat until the middle of the seventeenth century. A note in the parish register of Wensley, in Wensleydale, explains why there is no entry for the year 1563: 'the visitation or plague was most hot and fearful so that many fled and the town of

Wensleydale, the upper valley of the Ure, was originally river-named as 'Yoredale', but the present name was in use by about 1150. The market at Wensley was chartered in 1202, but the townlet was struck by the Black Death in 1563 and, when Askrigg gained a competing market in the same century, Wensley declined into the ranks of small villages

Wensley by reason of the sickness was unfrequented for a long season'. The trade of Wensley's market was captured by neighbours, and the settlement never recovered from this period of desertion. In such ghastly ways, the equilibrium between population and rural resources was restored.

Ecclesiastical and legal records provide the best sources of information regarding the impact of the disease. Thus it has been deduced that, in the archdeaconry of the West Riding, 45 per cent of vicars and rectors perished in the plague years of 1349–50; there was another significant outbreak in 1369, when Hampsthwaite, in Nidderdale, for example, which had lost its vicar to the plague in 1349, lost another. In the meantime, an eruption of the pestilence in 1361–2 had wiped out one third of the clergy in the deaneries of Richmond and Catterick. Jennings notes that: 'Of the 575 acres in Bilton with Harrogate, 274, or nearly 48%, was held by tenants whose deaths were recorded in the court roll of 1349–50.'

It appears that the disease had arrived in Hull in May 1349, reaching York towards the end of that month and afflicting most parts of the country in the course of the summer. During this terrible summer, 11 new graveyards were built to accommodate the bodies of victims in Yorkshire. In the winter of 1362, the chapel of ease at Stainburn was licensed for burials, since the roads were too deep in mud to allow corpses to be carried to the parish church at Kirkby Overblow, near Wetherby. Similar arrangements were made at several other places. Since the pestilence was carried by rat fleas, one might expect that it would wreak the greatest havoc in the filthy, congested cities and in the nucleated villages of the corn-growing areas, but be far less of a problem in areas of more dispersed population and pastoral farming. However, this argument is not borne out by the evidence from places like Nidderdale or the Calder valley, where mortality was severe, though a death rate of only 27 per cent has been calculated in remote Craven. Communities must have been shocked by the magnitude of the tragedy, and many survivors, bereft of friends and family and now living in crippled communities, must initially have envied the dead. Gradually,

however, they would have begun to appraise the nature of a situation in which the conditions of living and working had been transformed. Where recently there had been land hunger and overworked tenancies, there were now vacant holdings and abandoned fields. The labour market had been turned around; those who had been glad to sell their toil at minimal rates could now bargain and seek the highest bidder. Tenants who had been bound to their manors by threats and fines were now courted by lords who would gladly overlook desertions from a neighbouring estate. Authority struggled to enforce old rules and wage levels, but it became clear to all concerned that things could never be the same again. The old customs of servility were eroded more rapidly than before; the villein who gave his labour as boon work on the demesne was slowly superseded by the copyholder who paid rent and held land for life, or for several lives.

Gradually, tenants expanded their operations to take on the lands of former neighbours, or drifted into devastated estates where enticing terms might be offered. Wives and daughters often took over the holdings when their menfolk perished – like Alice Gagge of Harrogate, who took over her father's eight-'forest-acre' holding (covering about 30 acres). Thus, the number of villages extinguished by the Black Death in Yorkshire was probably very small indeed. However, scores of settlements would perish as indirect victims of the pestilence. It was true that peasants could now consider deserting manors where the lord was reactionary or the land was of poor quality, but lords were also tempted to rid themselves of what might have been regarded as headstrong and argumentative tenants. The aristocratic life did not depend upon mixed farming and the taking of rent, cash or services from peasants: wool prices were very high in the fifteenth century, and any landowner could expect to profit by running a large flock of sheep across his estate. The eviction of villagers across England as a whole was on such a large scale that the stability of the kingdom was put at risk and (largely ineffectual) commissions were appointed to investigate the situation. An enquiry in 1517 noted 84 instances of village destruction in

Yorkshire, involving the enclosure of 7,848 acres of land and the demolition of 232 homes.

Yorkshire is a notable graveyard of settlements, and the majority of the deserted medieval villages here may well have been victims of the Tudor sheep clearances. Most vulnerable would be those villages weakened by pestilence or desertion, with shrinking ploughlands and tempting pastures. Gristhwaite, in the North Riding, was a special case; probably weakened by Scottish raiding, it was depopulated and its lands enclosed. Close by was East Tanfield, where 17 families were taxed in 1301 and which had supported 14 taxpayers in 1332, but was deserted after its lands were enclosed and 32 people were evicted from the eight surviving houses destroyed early in the sixteenth century. Meanwhile, on the North York Moors, the village of Dale Town had more than 20 tenants in the fifteenth century, but, by 1569, sheep clearances had reduced it to 'a house called Dale Town'. The Tudor clearances had less impact in the dales, where villages were fewer and the pre-eminence of grazing was long established, though the weak community of Eavestone, near Pateley Bridge, was evicted around 1520. In the Wolds, parts of the Vale of York and flanks of the North York Moors, the effects were severe and whole tracts of countryside could be denuded of villages. Hey writes: 'On the high chalk Wolds the graziers were in their element. The large amount of ridge-and-furrow patterns revealed by aerial photographs confirms that the High Wolds were tilled extensively in the Middle Ages before the change to sheep farming.' Within a ten-mile radius of Richmond, at the junction of the Dales and Vale of York, there are about 30 deserted village sites.

As described above, Wharram Percy was once a substantial village. It was owned by the Percys and Chamberlains until 1254, when the Percys obtained the whole property. In 1402 they exchanged the village with the Hiltons for lands in Northumberland. By 1500, the Hiltons had completed the depopulation of Wharram and a sheep farm later appeared in its place, though the parish church survived to serve neighbouring communities for several centuries. Wawne, in north Humberside, had an uneven history. It appeared

in Norman times as a loose agglomeration of about a dozen dwellings. This undisciplined plan was abandoned in the late fourteenth century, perhaps as a consequence of the pestilence, and the village was rebuilt as a single string of houses set out beside a street. It did not survive long in its new form and, like so many other villages, was soon to be abandoned. East Lilling epitomises the themes raised in this chapter; it lay at the foot of the Howardian Hills, on gently sloping boulder clay about nine miles (14kms) north–north-east of York. Swan, Mackay and Jones describe how an unlocated Saxon settlement or settlements developed somewhere near to an east–west trackway linking Flaxton and West Lilling. In the early or middle twelfth century, a small, single-row, planned settlement of only five tofts, with field land and, perhaps, a fish-pond, was set out, possibly by the de Bulmer family. As its population grew, the settlement spread, gaining one toft to the east and two to the west. As the population grew further, at least four or five of the tofts were subdivided; dwellings were built in the new divisions and an extra fish-pond was dug. Land was in short supply, so the village street was diverted along a track or baulk and its former course was ploughed up. By the early fourteenth century, a decline of population had set in; by 1377, only 6–10 family tofts remained and some ploughland had gone out of use. In the fourteenth century, the deer park at Sheriff Hutton was extended across the westernmost part of the village and its fields and, by 1480, East Lilling was only a tiny hamlet of two or three households. The place was demolished and enclosed within the deer park by Richard III, as lord of the manor. Only the manor itself was preserved and developed as a farm. A park-keeper's lodge was built over the two central crofts and the park-keeper raised cattle on the deserted village site, cutting crew-yards for the beasts into the sides of the old street and ponds to the south of it.

Lost village landscapes vary enormously. If the site has been ploughed, then the only traces of a settlement may be the scatters of medieval pottery in the ploughsoil. If the land has remained under pasture, one may still recognise the corduroy patterns of the ridge-and-furrow ploughland, the

The best countrysides contain facets inherited from many different ages. In this view, looking down the Skyreholme valley towards Wharfedale, medieval strip lynchets pleat the ground in the middle distance; the hillock to their right looks like a Bronze Age barrow or stone clearance cairn. Several phases of enclosure have produced the field patterns; the commercial forestry is a modern addition to the scene and rest uneasily with the other components

trough or holloway of the old High Street, the boundaries of house plots, and the rectangular outlines of some dwellings. Occasionally, as at Wharram, Barford-on-Tees, or South Cowton in the North Riding, a deserted church may survive. The countrysides of Yorkshire are punctuated with the remains of deserted medieval villages. A high

proportion of these places fell victim to the Tudor sheep clearances, but a reconstruction of their biographies would find more than the popular theme of the cruel and rapacious feudal landlord. By the time that the last of their inhabitants were cast out onto the road, many villages had been fatally weakened by the overworking and exhaus-tion of their soils, undermined by the mortalities from the pestilence, and, sometimes also, scorched by the Scots. Worn out by their struggles against adversity, old villages took their rest beneath the sheep runs: some still lie there intact, awaiting a time when funds to excavate the historical heritage of common people become available.

REFERENCES

F. A. Aberg and A. N. Smith, 'Excavation at the medieval village of Boulby, Cleveland', in T. G. Manby (ed.), *Archaeology in Eastern Yorkshire* (Sheffield: 1988), pp. 176–88.

H. M. Beaumont, 'Tracing the evolution of an estate township: Barden in Upper Wharfedale', *The Local Historian*, 26 (1996), pp. 66–79.

M. W. Beresford and J. G. Hurst (eds) *Deserted Medieval Villages* (London: 1971).

M. W. Beresford and J. G. Hurst, *Wharram Percy* (London: 1990).

M. W. Beresford and J. K. St Joseph, *Medieval England: An Aerial Survey* (Cambridge: 1979).

S. D. Brooks, *A History of Grassington* (Clapham: 1979).

L. Cantor (ed.), *The English Medieval Landscape* (London: 1982).

R. A. Dodgshon and R. A. Butlin, *An Historical Geography of England and Wales* (London: 1978).

M. L. Faull and S. A. Moorhouse (eds), *West Yorkshire: An Archaeological Survey to AD 1500*, 4 vols (Wakefield: 1981).

R. Fieldhouse and B. Jennings, *History of Swaledale* (Chichester: 1978).

B. Harrison, 'New settlements in the North York Moors 1086–1340', in B. E. Vyner (ed.), *Medieval Rural Settlement in North-East England* (Durham: 1990), pp. 19–32.

D. Hey, *Yorkshire from AD 1000* (London: 1986).

W. G. Hoskins, 'Harvest and hunger', *The Listener* (1964), pp. 931–2.

J. G. Hurst, 'The Wharram research project: results to 1983', *Medieval Archaeology*, 28 (1984), pp. 77–111.

J. G. Hurst, 'Medieval rural settlement in eastern Yorkshire', in T. G. Manby (ed.), *Archaeology in Eastern Yorkshire* (Sheffield: 1988), pp. 110–24.

B. Jennings (ed.), *A History of Nidderdale* (Huddersfield: 1967).

B. Jennings (ed.), *A History of Harrogate and Knaresborough* (Huddersfield: 1970)

B. Jennings, 'Man and the landscape', in *The Calderdale Way* (1978).

J. McDonnell, 'Medieval assarting hamlets in Bilsdale, North East Yorkshire', *Northern History*, 22 (1986), pp. 269–79.

J. McDonnell, 'Pressures on Yorkshire Woodland in the later Middle Ages', *Northern History*, 28 (1992), pp. 110–25.

R. Muir, *Lost Villages of Britain* (London: 1982).

M. L. Parry, 'Secular climatic change and marginal agriculture', *Transactions of the Institute of British Geographers*, 64 (1975), pp. 1–14.

A. Raistrick, *The Pennine Dales* (London: 1968)

A. Raistrick, *Arthur Raistrick's Yorkshire Dales* (Clapham: 1991).

D. A. Spratt and B. J. D. Harrison, *The North York Moors* (London: 1989)

V. G. Swan, D. A. Mackay and B. E. A. Jones, 'East Lilling, North Yorkshire: the deserted medieval village reconsidered', *Yorkshire Archaeological Journal*, 62 (1990, pp. 91–109.

T. Williamson, *Polite Landscapes* (Baltimore: 1995).

CHAPTER THIRTEEN

Yorkshire Castles

ANY 'WORKING' CASTLE HAD TO EVOLVE in order to survive and, in Yorkshire, there exists a spectrum of castle types, making it possible to see how one form of defensive architecture was succeeded by another. In reviewing the types, one needs to remember that castles were built for different purposes: there were royal castles, built to buttress the royal interests in a region; private strongholds, built to protect a dynasty and its estates and to proclaim the importance of the owner; and other castles, usually royal, built to defend the state against foreign invasion threats. In medieval times, Yorkshire could be an unstable and potentially rebellious region; although well-insulated against continental invasion threats (with the Scandinavian challenge petering out in the early medieval centuries), it was vulnerable to Scottish raiding. However, it also had a highly strategic location, appreciated from at least the early Roman times, as a launching pad for Scottish campaigns and as a relatively secure supply base for armies campaigning in Scotland. Equally, Yorkshire lay astride the invasion routes into the English heartlands; the main bastions which faced the Scots were placed further north and, when Scottish raiders passed the castles of Northumberland and Durham, severe problems for the kingdom would result if the southward advance could not be stopped in Yorkshire.

After the Norman conquest, Yorkshire, like other parts of a virtually castle-free English kingdom, experienced a rash of castle-building projects as the king and his vassals sought to consolidate their grips upon the captured territories and,

The keep at Richmond Castle still symbolises the Norman conquest of Yorkshire

consciously or unconsciously, to demonstrate the ascendancy of a new culture. These castles were of two basic types: the now-familiar motte, or motte and bailey; and the little-known ring-works. Mottes were essentially a Norman introduction; although a few examples were built in Wales by Norman favourites of Edward the Confessor, they only became numerous after the conquest. Ring-works were circular, banked and ditched enclosures which resemble ring doughnuts when seen from the air, and there is some evidence that they could have had a slightly longer history than the motte. Yorkshire has a few examples of ring-works; one standing beside the early Norman church at Kippax was probably the defensive and military centre of the de Lacy estates. Castleton Castle, situated at the western end of the Esk valley and commanding the river and a north-south route across the North York Moors, is thought to have been built by Robert de Brus between 1100 and 1104 as the focus of an estate which encompassed 40 villages. It was built as a horseshoe-shaped ring-work with its entrance defended by a ditch. Unfortunately, our understanding of this under-researched class of defence works is constrained by a lack of modern excavations.

Mottes were flat-topped conical mounds of earth or rubble which were normally crowned with a palisade of stakes ringing a tall wooden tower. Often, an outer defensive enclosure, a bailey, guarded the main buildings of the lord, and the motte was defended as a refuge of last resort if the bailey was overrun. Good examples of such earthwork castles are Sandal Castle, near Royston, and at Laughton en le Morthen, near Rotherham. The motte usually stood to one side of the bailey, but at Barwick in Elmet the mound is at the centre of the bailey, while the motte and bailey together occupy just a corner of an Iron Age hill-fort. Almondbury hill-fort was also exploited as a Norman manorial centre, and the Paynel manor at Bingley is thought to have been administered from a site inside the ancient earthworks on Bailey Hills. At Skipsea, now in Humberside, the arrangement was quite unusual, with the motte some distance from the bailey and linked to it by a causeway spanning the tidal mere which lay between them. At Burton in Lonsdale, a ring-work was converted into a motte and bailey castle,

leading to suggestions that ring-works may have been favoured for their expediency of construction in territory considered to be hostile. In Yorkshire, the Norman earthwork castles were the administrative capitals and refuges of unpopular masters in a land that was resentful and unruly – but they were adequate to the task of stamping their owner's authority on the disorganised rural communities. More general uprisings exposed the vulnerability of their timber fortifications, and both the royal mottes in York, built on either side of the Ouse, were sacked in 1069, when the local people joined forces with a Viking raiding party. (This weakness was demonstrated, again in a shameful manner, in 1190, when a York mob attacked the Jewish community of the city, which sought refuge in the wooden precursor of Clifford's Tower.) There was, however, a question of costs to be considered: in 1192, a timber keep was built at York for £28 13s. 9d., while the stone replacement cost £1927 8s. 7d. (the surviving stone keep, Clifford's Tower, which crowns the old motte, has a quatrefoil plan of four semi-circular bastions and dates from 1245).

As the Norman period advanced, the old earth and timber strongholds either became obsolete, purely administrative centres, or they were superseded by costly stone-walled castles. Ryder notes that:

A few castles, such as Richmond, possessed a stone curtain wall from a relatively early date; as the 12th century proceeded, and the Anarchy of Stephen's reign passed, earthwork castles were either abandoned, or their defences remodelled in stone. The appearance of the Norman great towers or keeps (Richmond c. 1150–70, Newcastle c. 1168–78 etc) introduced a genuinely defensible unit into perimeter-defence sites, although in practice retreat to the keep seems to have been very much a last resort; the strength of most castles lay in their ditch and curtain wall, a strength emphasised by the 13th century trend of adding mural towers and rebuilding gatehouses.

Occasionally, as with Clifford's Tower, a stone keep was built upon the old motte mound, although the subsidence of the earthwork under a great weight of masonry was often recognised as a threat to such endeavours. In such cases, castle builders had other options: to ring the motte with walls to create a 'shell keep'; to build a stone tower keep on stable ground; to circle a ring of bailey defences with a wall to create a curtain; or to combine these different options. Pickering has an excellent example of a 'shell keep' of the 1220s. The original castle, built during the reign of William the Conqueror, was a motte-and-bailey earthwork; stone fortifications were added after 1180. Ryder interprets the defences as follows: 'At the centre of castles like Mitford and Pickering the shell keep crowning the motte was in essence a group of buildings arranged around a small courtyard and backed by a curtain wall, a scaled-down version of the perimeter-defence concept and possibly the genesis of the later medieval "concentric" castles [e.g. Beaumaris or Harlech] so well seen in the result of Edward I's Welsh campaigns.' Pickering was the administrative centre of a large honour and the headquarters of the royal Forest of Pickering. Since square towers, with their vulnerable corners and weight of masonry, could present problems, the motte there was crowned with a circular wall, from which wing-walls ran down the sides of the motte to link up with the curtain of wall around the bailey.

The early tower keeps had a rectangular plan which rendered them liable to collapse if sappers were able to undermine the masonry converging in their corners. An advanced and impressive remedy to this problem can be seen at Conisbrough Castle, in the south of the region, where earthwork defences had been built by William de Warenne after the conquest. The keep, dating from 1185–90, is cylindrical. It stands on a splayed base which was intended to frustrate attacks on

The motte-and-bailey earthworks at Pickering date from the reign of William the Conqueror. A stone shell keep was built to ring the summit of the motte at the end of the twelfth century, when a curtain wall with interval towers was added to protect the domestic buildings and castle which had developed around the motte. A fragment of the shell keep can be seen on top of the motte, between the two trees in the background

Conisborough Castle, where the cylindrical keep dates from 1185–90

the foot of the wall, while six solid buttresses project from the curving walls. From the tops of these buttresses, defenders could sweep the walls with fire; sappers working at the base of the tower would have been particularly vulnerable. Even before they could attack the keep, they would have had to breach the curtain wall, 10.7 metres high and 2.1 metres thick, which surrounded the inner bailey. The towers in the curtain wall here are of an early and primitive form, being solid drums of masonry reminiscent of the bastions on Roman forts.

Richmond Castle is more traditional in appearance and represents one of the finest examples of a rectangular keep. At Richmond, apparently, the bailey was enclosed by stone walls from the outset, one of the earliest examples of stone curtain wall appearing here after 1071. A steep cliff formed the defences on one side of the triangular enclosure, the stone curtains guarding the other two sides. There was a gateway and a gate tower in the apex of these walls, and, in the second half of the twelfth century, the gate tower was used as the base for the 15.8 metre stone keep, which is still the commanding feature of the

attractive town. After the blocking of the gate, the keep was entered at the second-storey level, from the rampart walk of the curtain tower. The tower is remarkably well-preserved, standing to its original height and maintaining the battlements of the corner towers. Hey notes that:

> The honour of Richmond was one of the three largest feudal holdings that William created, with 440 dependent manors in many parts of England, the Yorkshire section of the lordship formed an unusually large and coherent estate of 199 manors and 43 outlying properties, which together established a powerful military presence in the north-western part of the county near the junction of the main routes coming out of Scotland into the northern Vale of York. Richmond was granted to Alan the Red, or Alan Rufus, a cousin of the Count of Brittany, and it continued with this Breton family until 1399.

The North Riding boasts another great square keep in the former estates of Alan the Red, Middleham Castle, which is less well-preserved than the castle at Richmond and of a different design. Originally, 'William's Hill', a motte-and-bailey castle, guarded estates and roads in Wensleydale; this arrangement was abandoned when work on a new castle nearby began in about 1180, and the remains of the earthwork castle can be seen on the rising ground behind its stone successor. The oblong keep was very large and massive, so that the space which would usually have served as the great hall was divided into two rooms, it being too big to roof with a single span of timbers. The keep was surrounded by a broad alley-way, beyond which lay other buildings, arranged so that their outer walls were integrated into a square curtain wall which was added in the thirteenth century. With proportions that were squat rather than lofty, Middleham was a castle of a type known as a 'hall keep', a design favoured by those who were prepared to surrender a small measure of military excellence in order to enjoy the domestic comforts associated with more spacious chambers and fewer stairs. Brown notes that: 'Middleham Castle must have reached some importance by the turn of the twelfth century for not only is it mentioned in Gervase of Canterbury's "Mappa Mundi" (1200), but King John tried to place it in the hands of one of his supporters, Nicholas Stapleton, in 1216. This attempt failed but it shows that Middleham was beginning to emerge as a centre of government in the north.'

Scarborough Castle was built on Castle Dyke, a steep ridge at the end of a headland which was created by faulting and composed of calcareous grit and coralline oolite. This naturally defensible site was intermittently occupied or defended from Bronze Age to Viking times; Farmer notes that in 1066: 'a settlement large enough to offer substantial resistance to the landing of a Scandinavian force under Harald Hardrada and Tostig had become established'. The remains of a Roman signal station lie nearby. The strategic importance of the position overlooking the east coast must have been valued; the medieval castle began as an unlicensed or adulterine castle which was built during the anarchy of Stephen's reign by the local magnate, William le Gros, Count of Aumale, Earl of Albermarle and York, and the chronicler of Meaux Abbey recorded that William: 'observing this place to be admirably suited for the erection of a castle, increased the great natural strength of it by a very costly work, having enclosed all the plain upon the rock by a wall and built a tower at

the entrance'. This castle must have resembled the one at Whitby, but when central authority was re-established under Henry II, who advanced on the North with his army in 1155, the Count of Aumale surrendered Scarborough and the castle that he had seized at Pickering. Rather than slighting Scarborough Castle, the normal fate for adulterine strongholds, Henry appreciated its strength and significance and redeveloped the site as a royal castle, spending £532 on the works here between 1157 and 1164. Henry's keep was completed in 1169, but his son John spent the great sum of £2,000 on improving the castle and was probably responsible for furnishing the curtain walls with semicircular towers. In fact, John spent more on his Yorkshire royal castles at Scarborough and Knaresborough than on any others in the realm, though he was more troubled by fears of his barons than by Scottish invasion. After the Civil War, the west wall of the keep at Scarborough was blown up by the Parliamentarians; troops were billeted in barracks built in the outer bailey after the Jacobite invasion of 1745; the ruins were shelled by German cruisers in 1914. Most keeps had three floors, but Scarborough Castle was a towering building with four storeys. Bowes Castle, commanding the route over Stainmore Pass and standing in a corner of the Roman fort of *Lavatrae*, was a royal hall keep of more modest dimensions. It was built by Richard in the 1170s and 1180s to a rather antiquated design, and lacked the towered curtain wall generally considered necessary by this time to keep sappers away from the base of the keep.

Other notable castles of this period, now however, more ruined, included Pontefract Castle and the one at Knaresborough; and barons who revolted against King John included the former holders of these confiscated castles, Lacy and Stuteville respectively. Both castles were largely dismantled during the Civil War. Ryder observes that: 'At Alnwick, Pontefract, Sandal and Skipton the 13th century saw the combination of the shell keep and mural tower concepts produce complex clustered donjons, a true castle-at-the-heart-of-a-castle.' Sometimes it was decided to give priority to the curtain wall, resulting in the appearance of 'castles of eincente', or castles of enclosure. Kilton Castle, near Saltburn, was rebuilt

Helmsley Castle, with the old keep tower undergoing repairs in the background. The massive D-shaped towers, built in about 1250, guard the entrance, anticipating the emphasis that would come to be placed on gatehouse fortifications

in this manner around 1190, but is now very ruined. It appears that Pagan Fitz-Walter was obliged to flee from this base at Seaton during the anarchy and, in about 1140, he founded the castle at Kilton as a timber stockade on a steep-sided promontory which could be isolated by the cutting of a deep ditch; later in the twelfth century the timber defences were replaced by stone. As the black arts of war developed, attention tended to focus on strengthening the vulnerable entrances in the curtain wall. Periodically, a castle would fall to treachery; a keep or gatehouse tower which could serve as an ultimate stronghold was valued as a bolt-hole to which loyal defenders could retreat in the event of a revolt by the garrison or a breaching of the curtain. It was realised that several needs could be reconciled by the building of formidable gatehouse towers, while increasing strength and sophistication in design were prompted by increasing sophistication in offensive military engineering. Different responses to the challenges are represented by the castles at Helmsley and Skipton. At Helmsley, the town is overlooked by the slab-like face of the ruined

keep. The recently restored ruins reveal an interesting experiment, for when it was built, within the concentric earthworks of an earlier castle, by Robert de Roos in about 1200, the tower was designed to a D-shaped plan, with a curving outer face. The keep is attached to the surrounding curtain wall of the bailey, which was furnished with projecting circular corner towers, and the northern entrance was guarded by two drawbridges and two great D-shaped towers, the D-form being a similar improvement on rectangular towers as the circular keep on some of the weaknesses of the square form. The integration of the keep into the curtain wall has been praised by experts in military architecture; Illingworth wrote that 'Robert's curtain is an excellent example of the best practice of the time.' Barbicans were added to the two gateways in the middle of the thirteenth century, with the great barbican at the south of the castle being linked to the curtain walls by wing-walls; the emphasis throughout was on mobility and accessibility within the castle. Illingworth comments that: 'The effect of this easy access to all parts of the fortifications was greatly to increase the mobility and effectiveness of a small garrison and to make hazardous the work of an enemy attempting a direct attack on the curtain. The two gates, one at either end of the enclosure, held over the besiegers the constant threat of a sortie and attack on flank or rear.' At the close of the thirteenth century, the keep and the curtain walls were raised, probably to counter improvements in the design of siege-engines.

At Skipton, the Norman castle overlooked the settlement from a nearby knoll. It was developed into a formidable stronghold in the early fourteenth century, with a small bailey guarded by a cliff on one side and enclosed by a semicircular curtain studded with six massive circular towers; its strength was concentrated in the massive towers which flanked the inner gatehouse, while an outer gatehouse, with four strong towers, guarded the entrance to the outer bailey. Skipton remained the stronghold of the Clifford family for centuries to follow. It was partly demolished and rebuilt by Lady Anne Clifford in the sixteenth century, and the octagonal tower is an addition of 1536. These changes symbolised the evolving role of the castle, and Skipton was remodelled to meet the needs of comfort and conformity rather than defence. At Helmsley, the military phase in the development of the architecture ended at the close of the thirteenth century with the raising of the curtain and keep; thereafter, additions were concerned with increasing comforts and amenities. A great hall was added in the late fourteenth century, along with a range of domestic offices. The heyday of the English castle was the fourteenth century, though developments were already foreshadowing the decline of the castle – while nobles still responded to the prestige attached to towers and barbicans, the emphasis shifted from substance to style. Then, with changes in the conduct of warfare, the centralisation of power and the banning of private armies by Henry VII, there was little future for the dynastic stronghold, though plenty of scope for gracious living. Cathcart King noted various differences between northern and southern preferences in castle design, and added:

Another difference between north and south is the opulent character of many northern castles of the fourteenth century, particularly those of the quadrangular type. The typical southern castle of the period was built by the head of an emerging family – not infrequently by a successful soldier like Dalyngrigge [of Bodiam Castle] – and had modest, though comfortable and up-to-date domestic accommodation.

Many northern castles, on the other hand, were robber-proof palaces built by established nobles; over most of England buildings of this kind would never have needed so much as a prestence of fortification, but these northern examples are not simply fortified by way of pretence. On the whole they are weaker than their opposite numbers in the south – certainly relative to their domestic splendours. In any case, the southern castles are superior in terms of military science.

Bolton-in-Wensleydale is the best surviving example of a quadrangular castle of the fourteenth century. Richard, Lord Scrope, obtained his licence to crenellate or fortify in 1378, and the castle that he built had four square corner towers, linked by ranges of accommodation to surround a quadrangle or oblong courtyard. In defensive terms, the castle had distinct limitations, but it did provide eight different suites of accommodation to facilitate the hospitality which was expected of a noble of this time. The site has no military advantage and it is overlooked; it was chosen purely for convenience, for the northern and western ranges of the castle were built beside the manor house of Old Scrope, which was then demolished to make way for the remainder of the castle. Gardens were planted on the terraced slope beside the castle. It may have been at the time that the castle was built that the adjacent, neatly planned village appeared, although the two are slightly out of alignment, so the village could have come first. Cathcart King appraised the advantages and disadvantages of Castle Bolton as a defence work:

Its four big square corner-towers have only a very shallow projection, and the weakness of its flanking defence is hardly cured by the two little turrets in the middle of the longer curtains of its rectangular plan. There is no wet moat, and very little of any sort of ditch at all; the loops on the outer walls are not real arrowslits, and there are no gunholes. But the huge mass of building – three storeys high with the corner towers rising another two storeys above the curtains – has a massive strength; the walls are of respectable thickness,

Castle Bolton, in Wensleydale, symbolises the gradual transition from castle to country mansion

and the entry has a singular feature: the gate itself has two portcullises, and in addition every doorway into the buildings surrounding the courtyard was protected by a portcullis. The sides of the quadrangle are almost as grim as the outer walls. There was nothing perfunctory about the defences of Bolton.

Castle Bolton encapsulates the changing aristocratic outlooks of the fourteenth century and, despite its military weaknesses, it did hold out against a siege by Parliamentarians for some months in 1645, although the royal garrison eventually surrendered. (A large proportion of Yorkshire castles were defended during the Civil War and, as a result, many suffered bombardment and slighting – were it not for the Civil War, much more medieval defensive architecture would survive. Pontefract, Helmsley, Scarborough and some others were severely damaged.). A rather similar castle of the quadrangular palace-castle type, also a building of the late fourteenth century,

stands in ruins at Sheriff Hutton, near Malton. In its day, it must have been an even more imposing example of the genre than Bolton. Ripley Castle, near Ripon, has a fifteenth-century gatehouse, a sixteenth-century tower and various eighteenth-century embellishments, and takes the story of the military decline of the castle some stages further.

Castles were not the only fortifications to be seen in medieval Yorkshire; different ends of the defensive spectrum were represented by the fortified house and the town wall. Superficially fortified houses existed in most parts of medieval England; in the Midlands and East Anglia, moated homesteads, usually manor houses, were exceedingly common. Some 320 moated sites are recorded in Yorkshire, where they cluster in the lowland areas with high water-tables, notably the Vale of York and Holderness. They were largely built during the century following 1250, mainly to surround manor houses, though some were associated with granges and other monastic properties. However, while the moats might have deterred some local ruffians, they were principally status symbols, underlining the claims to noble status of the family living inside the moat. Some Yorkshire houses had more purposeful defences, the main cause of insecurity being the Scottish problem. At the time of the Norman conquest, there was no clear-cut divide between the people of northern England and southern Scotland. In both areas, there were populations with a mixture of British, English and Scandinavian cultures; Scottish nationhood had scarcely developed, while Yorkshire and the north-eastern counties might have fitted into Scotland as easily as into England. Naturally, the Norman kings could not countenance any strengthening of northern ties with Scotland and they recognised that York, then the second city of the realm, was highly vulnerable to any Scottish invasion force which could enter the North Riding. To prevent this happening, a royal castle was established by the Tyne at Newcastle; the Bishop of Durham was given

great military powers within his county and, in time, the dynasties of Percys, Nevilles and Cliffords were allowed to wield regal powers in their frontier provinces. Thus Yorkshire was cushioned and protected by the buffer statelets and castles of the marchlands to the north.

However, royal power in the North of England virtually collapsed during the protracted civil wars of King Stephen's reign – and the Scots invaded. An independent alliance of forces raised by the Archbishop of York and the northern barons was hastily formed, and the Scots were routed at the Battle of the Standard, fought near Northallerton in 1138. In the decades that followed, Scottish raiding parties periodically penetrated England, causing massive reprisals by the English, but the

The romantic ruins of the castle at Sheriff Hutton. Only the broken corner towers of the 'quadrangular palace castle' of 1382 still stand; originally, there would have been a strong resemblance to Castle Bolton

Makenfield Hall, a superb example of a defended medieval house

---

English defeat at Bannockburn in 1314 was the trigger for large-scale Scottish invasions and there were English defeats at Myton-on-Swale in 1319 and Byland in 1322. The town of Knaresborough was protected by a ditch that was linked to the deep, dry ditch of the castle. However, Jennings notes that: 'The rampart and ditch offered no protection in 1318, when Knaresborough was attacked by the Scottish raiders. According to the official record, 140 of the 160 houses in the town were burned down . . . The townsmen may have ceased to maintain the "town dyke" before the Scottish raids. It was being encroached upon for building and agricultural purposes from the middle of the fourteenth century.' Then the tides of war turned, and the Scots were defeated at Dupplin Moor in 1332, Halidon Hill in 1333 (where the English longbow first demonstrated its ruthless efficiency), Neville's Cross in 1346 and, crushingly, at Flodden in 1513. In the meantime, estates and monastic houses had been

plundered by the invaders and Knaresborough was burned.

The long history of raiding naturally gave rise to feelings of insecurity in Yorkshire, as well as to deep-seated beliefs that southerners were not greatly concerned about the safety of their northern compatriots. However, when one looks at the relics of supposedly fortified houses, it is not always clear where fortification ends and where decoration begins. Also, the categories of castle, tower house, fortified house and pele tower tend to overlap and blur. It is safe to say that some features were genuinely defensive, although in many cases the owners would have been well advised to do as the lesser folk did – and take to the woods when serious raiding threatened. Ryder comments that:

On a lower social level, in some lowland parts of Yorkshire moat construction does not seem to have been wholly a defensive exercise; a moat probably conferred a certain status to a minor manorhouse. It might also serve to drain a boggy site, as well as providing a useful barrier to keep domestic animals in and wild ones out. Of course, on most sites defensive features combined an element of practicality with an element of status; quantifying each element now may be difficult.

The families of the lesser Norman lords would frequently set up home in the modest shelter provided by a small motte. A couple of centuries later, their descendants were likely to have been found living in a moated, timber-framed homestead, and it is probable that about 100 more of these homestead moats were dug in medieval Yorkshire than the 320 recorded examples. Ryder speculates that: 'Moated sites can be seen as the natural successors to the smaller earthwork castles of the late 11th and 12th centuries, as baileys which have lost the need for a motte ...'. Sometimes the earthworks of mottes and homestead moats can be found lying side by side, as at Topcliffe, near Thirsk. Some of the grander of the moated and fortified houses were built in stone; one of the finest survivors in the whole of England is Markenfield Hall, near Ripon. John de Markenfield obtained his licence to crenellate from Edward II in 1310; as Chancellor of the Exchequer, he was an important public figure. Defensive considerations were handled thoughtfully; inside

the rectangular moat, the buildings are arranged with their backs to the water around a courtyard and the house was built with the great hall placed at first-floor level, echoing the quasi-defensive tradition of some much earlier Norman stone houses. There were originally an outer wall and moat, with an outer gatehouse placed on the approach from the Ripon–Ripley road. A tall chapel was attached to the south wall of the house, where the remainder of the buildings were probably connected with farming and storage. The house would have been incomplete or newly built when the Scots burned Knaresborough in 1318; it escaped, but it would be interesting to know if its defences were tested. The house was forfeit to Elizabeth I after the Rising of the North and she sold it to her Lord Keeper, Sir Thomas Egerton, who altered it and added the surviving 'gatehouse', really an ornamental porter's lodge. Spofforth Castle is very different in appearance. The first licence to crenellate was obtained in 1308; the building work must have taken place

Nappa Hall, in Wensleydale, a pele tower of the mid-fifteenth century

above the basement of the earlier thirteenth-century house, which is dug into the rock of the bluff. Two fourteenth-century bays remain, although the great hall is a product of rebuilding in the fifteenth century. This was a seat of the Percy family, and little survives to suggest that it was seriously fortified; the building was probably more of a palace than a castle.

Tower houses and lesser pele towers of various sizes were numerous in the lands further north, where the Scottish threat was more immediate and the barons more unruly. A smaller number of examples were built in Yorkshire, some of which have survived. Hellifield Peel was quite a massive stone tower of the mid-fifteenth century which experienced some changes of detail in the

seventeenth and eighteenth centuries. Cowton Castle, near Richmond, is of a similar date; both would have offered some protection against smaller bodies of raiders. An alternative to the tower-house form involved attaching a tower to the domestic accommodation. One of Yorkshire's best examples of a pele tower is Nappa Hall, in Wensleydale, where the domestic accommodation is sandwiched between a greater and a lesser tower, with the tall west tower being the principal fortified bolt-hole. Nappa Hall dates from 1450–9 and the chronicler, Leland, described how Thomas Metcalfe, who had fought at Agincourt in 1415, 'waxed rich and builded the two faire towers'. At Mortham Tower, in Teesdale, the fourteenth-century tower stands amongst later additions to the house; Bolling Hall in Bradford has also survived in a changed context. Other interesting relics include the gatehouse tower of the Marmion family, which stands beside the church at West Tanfield, and the octagonal brick tower at Temple Manor, Temple Hirst, created during the rebuilding of a preceptory of the Knights Templars. Almost all the fortified hunting lodges have

---

The church tower at West Tanfield; the building immediately to its right is the gatehouse tower of the Marmion dynasty. Since its defences could have been swept clean of defenders by any attackers who gained control of the overlooking church tower, the Marmion Tower is best regarded as a status symbol

disappeared, although Barden Tower in Wharfedale is still imposing. Fragments of a late medieval hunting lodge form Norton Tower on Rylstone Fell, and Haverah Park contains the ruins of 'John of Gaunt's Castle'.

While the lords had their castles and tower houses, the ordinary rural people were undefended. When trouble threatened, a few members of the lower orders and some monks and priests might find shelter in a local castle, but people normally dispersed to the more sheltered corners of the countryside until the raiders had passed by. Some church towers are said to have served as refuges. The newel staircase in Bedale Tower was furnished with a portcullis, and the towers at Masham, Middleham and Thornton Watlass had upper chambers provided with fireplaces. Tadcaster contained one of the various churches burned by Scottish raiders after the English defeat at Bannockburn.

Town walls constitute defences on an incomparably grander scale, and there are no better examples than the medieval walls of York. The actual function of town walls is debatable; in

The tower of Bedale church is thought to have served as a refuge and its staircase was protected by a portcullis

some cases, the walls were a definite defensive asset, though the longer the walls of a town were, the more difficult they were to man and defend properly. In England, to a greater extent than on the Continent, town walls were used to signify the importance of a place, and civic pride has a long history. In addition, the possession of walls with a limited number of gateways made it much easier to regulate access to markets and to collect appropriate tolls. On the other hand, walls could act as a corset on urban growth, leading to congestion and forcing the expansion of lower-grade development in the extra-mural area. At their maximum extent, the walls of York ran for more than two miles, and the greater part of this circuit has survived. Little Roman fabric remains, but the medieval walls trace the line of the walls of the Roman legionary fortress in the north and west, though the total area enclosed is about four times

that of the fort. Repairs to the Roman circuit were attempted by Danish forces in 876. The early medieval walls were of an earth and timber construction, although stone gateways may have been built in the eleventh and twelfth centuries. The stone walls date from the thirteenth and fourteenth centuries, and the four gates (Bootham Bar, Micklegate Bar, Monk's Bar and Walmgate Bar) were rebuilt in the mid-fourteenth century. Only Walmgate Bar still preserves its outer gate or barbican; the other barbicans were destroyed in the nineteenth century. Bootham Bar is the oldest of the gatehouses, retaining a Norman arch of the late eleventh century. Until the gatehouse at Monk's Bar was built in the fourteenth century, the Roman gate may have remained in use. In addition to the gatehouses, the medieval walls had 6 smaller gates or 'posterns', exemplified by the surviving Fishergate, and 44 towers, of which 39 survive. Two towers, the Anglian and the Multangular, have complex histories, the Multangular Tower having fourteenth-century masonry crowning its Roman base; the D-plan towers are

Walmgate Bar, in the medieval walls of York, is the only gateway to retain its outer gate or barbican

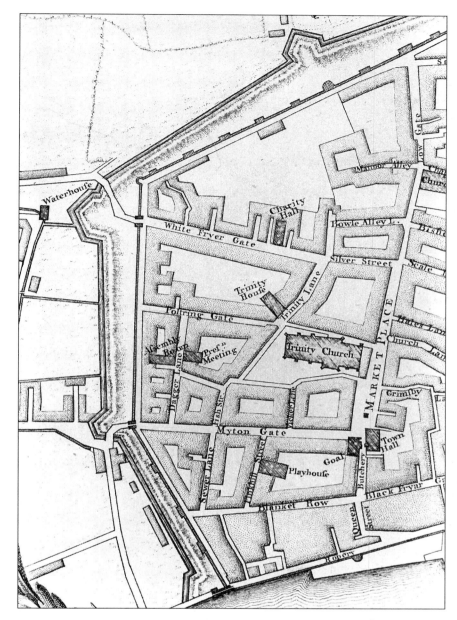

Jeffrey's map of Hull in the late eighteenth century, showing the medieval walls and earthen artillery defences of the Civil War (now lost)

probably of the thirteenth century, the rectangular ones are of the fourteenth century, and the half-hexagonal towers are of the late fourteenth and fifteenth centuries.

York was not the only defended town in Yorkshire, but it is the only one to have preserved its defences. Medieval Richmond was originally confined by a wall equivalent to an outer bailey of the castle. Beverley was not walled, but the medieval town was defended by a great ditch which restricted access to five places, controlled by brick gatehouses. Only one of these has survived – the North Bar, a superb building made of local bricks and built in 1409. Scarborough and Knaresborough were also defended by ditches. Kingston upon Hull, developed by Edward I after 1293, probably had four gates, which were almost certainly of brick, as well as defences which incorporated 25 brick towers, and the approaches of the town were guarded by an artillery fort, the Citadel. This was not the end of the story of fortification at Hull; in the Civil War, Hull was one of the towns which, like Newcastle, King's Lynn, Colchester and others, gained a circuit of artillery fortifications – a belt of earthworks with artillery positions hastily cast up, once impressive, but now vanished. Pontefract was another Yorkshire town defended in this way, with the defended area, which had the old castle at its centre, shielded by defences which linked up a series of forts, batteries and hornworks.

REFERENCES

F. A. Aberg (ed.), *Medieval Moated Sites*, Council for British Archaeology Research Report 17 (London: 1978).

J. Birch and P. Ryder, 'New Hall, Darfield, Yorks', *Archaeological Journal*, 54 (1982), pp. 81–98.

J. Birch and P. Ryder, 'Hatfield Manor House, South Yorkshire', *Yorkshire Archaeological Journal*, 60 (1988), pp. 65–104.

S. V. Brown, *Castle, Kings and Horses* (Middleham: undated).

D. J. Cathcart King, *The Castle in England and Wales* (London: 1988).

P. G. Farmer, 'Early medieval settlement in Scarborough', in T. G. Manby (ed.), *Archaeology in Eastern Yorkshire* (Sheffield: 1988), pp. 124–48.

D. Hey, *Yorkshire from AD 1000* (Harlow: 1986).

J. L. Illingworth, *Yorkshire's Ruined Castles*, 2nd edn (Wakefield: 1970).

B. Jennings, *A History of Harrogate and Knaresborough* (Huddersfield: 1970).

H. E. J. Le Patourel, *The Moated Sites of Yorkshire*, Society for Medieval Archaeology Monograph 5 (1973).

J. S. Miller, 'Restoration work at Markenfield Hall, 1981–4', *Yorkshire Archaeological Journal*, 57 (1985), pp. 101–10.

P. Ryder, 'Fortified medieval and sub-medieval buildings in the North-East of England', in B. E. Vyner (ed.), *Medieval Rural Settlement in North-East England* (Durham: 1990), pp. 127–40..

P. Ryder and J. Birch, 'Hellifield Peel – a North Yorkshire tower house', *Yorkshire Archaeological Journal*, 55 (1983), pp. 73–94.

S. J. Sherlock, 'Excavations at Castle Hill, Castleton, North Yorkshire', *Yorkshire Archaeological Journal*, 64 (1992), pp. 41–7.

B. Vyner (ed.), *Medieval Rural Settlement in North-East England* (Durham: 1990), pp. 127–40.

P. R. Wilson, 'Excavations at Helmsley Castle', *Yorkshire Archaeological Journal*, 61 (1989), pp. 29–33.

CHAPTER FOURTEEN

Medieval Churches of Yorkshire

YORKSHIRE CONTAINS A REMARKABLE ASSEM-BLAGE of religious buildings. The conversion of the region and the acquisition of a legacy of Saxon churches have been explored above. There are more than 600 Saxon and medieval churches in Yorkshire, and the great majority of these had been founded by 1200. It would be hard to recognise a Yorkshire style in Saxon architecture; as the population of churches was augmented and evolved, so the main European styles of architecture – the Norman, Transitional, Early English, Decorated and Perpendicular – found expression in the region's churches. At the same time, however, several Yorkshire idioms gradually began to emerge in different subregions. The regional influence was not simply determined by variations in the availability of building materials; economic

and cultural factors also exerted an effect. If a more general Yorkshire character found expression in the churches, it might be identified with the quest to 'get the basics right and keep the frills in check'. Yorkshire contains more than 500 medieval churches, most of them embodying several styles of building. The Norman conquest did not prove to be an immediate watershed in architectural history, even though the small, light and lofty style of the Saxon church contrasted with the

The nave arcade in Stonegrave church is Norman, but late Saxon features also survive. The wheel-headed cross, probably as late as the tenth century, suggests that Celtic influences persisted here long after the synod of Whitby

monolithic structure and distinctive decoration of the Norman version of the Romanesque. For some decades after the conquest, church building in Yorkshire was accomplished by Saxon masons working in the traditional manner, so that churches of the late eleventh and early twelfth centuries could be virtually indistinguishable from pre-conquest buildings: the church at Weaverthorpe, near Malton, for example, is Saxon in style but a product of the early twelfth century. The devastation caused by the Harrying meant that there was a lull in church building in the years that followed, but then, as efforts were made to rehabilitate the countryside, a vigorous programme of work was launched, particularly in the Wolds, so that scores of medieval churches have developed from a Norman nucleus. The Norman

enthusiasm for church building probably involved more than religious zeal, for churches built and decorated in the Norman manner must have existed as symbols of political domination.

Some of these churches were large, cruciform buildings with central towers, like the ones at Bossall, near York, and North Newbald, near Market Weighton. Others, more numerous, were simple nave-and-chancel buildings which were likely to be engulfed by later building work as congregations swelled. Adel, near Leeds, is a well-preserved little Norman church with unusually fine decoration around the south door and chancel arch. Some of the finest Norman architecture in the region was accomplished at Fishlake, near Doncaster, where elaborate medallions were carved around the doorway. In other places, the Norman period may be represented by fragments of incorporated masonry and decoration, apparently undatable stretches of wall with windows inserted later, or by surviving Norman fonts, like the ones at Bessingby, near Bridlington, or Nafferton, near Driffield. The Norman enthusiasm for apsidal chancels found only modest support in England, and there are very few examples in

The arrival of the Normans in Yorkshire is represented here by two unusually early Norman fonts. The one from Easby church (left) *is decorated with refined Romanesque carving, but the Burnsall font* (right) *shows the persistence of Anglo-Danish influences in the vigorous and rather 'primitive' decoration*

Yorkshire; the apse at Birkin, near Selby, is said to be the best. Norman towers were often heightened at later medieval dates, as at Birkin and at Wath upon Dearne, near Rotherham, where the spire dates from the fifteenth century. A superb Norman tower was added to the earlier Norman church at Campsall, near Doncaster. The shift from the Romanesque to the Gothic style is marked by Transitional work, like the arcading in the church at Wadworth, near Doncaster, the naves at Thorpe Salvin, near Rotherham, the great monastic church at Selby, and the tower at Drax.

One of the most remarkable items of Norman church building in Yorkshire does not consist of stone, but of timber and iron. In 1975, specialists in several fields of scientific expertise carried out an analysis of the south door of St Helen's Church, Stillingfleet, where the decorative ironwork includes a representation of a Viking-type ship. Bradley writes that: 'The study confirms the rarity, and in some points the uniqueness, of this door which, the authors of the published descriptions conclude, is best regarded as coeval with the doorway surrounding it, thus belonging perhaps to the third quarter of the 12th century.' Other representations, some fragmentary, appear on the door – a man and woman, a solitary man, a tree and a cross – and Bradley speculated about the nature of the iconography that had been created to influence all who passed the church door. It was suggested that the ship might represent the ark, the figures Adam and Eve, and the composition would be full of allegorical meaning:

Seen in these terms, the Stillingfleet door features a symbolic composition designed to affirm in simple images the cyclic pattern of divine intention perceived by Christian philosophers in the history of the world. According to this pattern, embodied in the popular formulation of the Legend of the Holy Rood Tree, the redemptive purpose of God is seen repeatedly prefigured in history from the moment of the Fall and reaffirmed in post-Incarnation history. Through the Church, by the waters of baptism and the way of the Cross, the righteous, the congregation of God, may assuredly achieve salvation.

With the establishment of the first Gothic style, the Early English, around 1150, it became possible to build churches that were larger, loftier and, in both senses of the word, lighter. The new possibilities were avidly explored in the Cistercian abbeys, where the change-over is often plainly preserved in the masonry. In the parish churches of Yorkshire, where the scope was more limited, the changes were less dramatic and simplicity was often the keynote. One small but fine exception is the church at Skelton, near York; built around 1240, it seems to have been the creation of the same masons who worked on York Minster. More frequently, the Early English architecture was modified and augmented by the products of later medieval building phases, as at Hemingbrough, Bossall, near York, and St Oswald, Filey. Similarly, the Decorated style of about 1250–1350 found its fullest expression in monastic buildings like Easby Abbey, and great churches like York Minster and Beverley Minster. One glorious exception to this rule is the church at Patrington, in Holderness, where the village was a bustling market town in medieval times and where the manor was held by powerful patrons, the Archbishops of York. Almost entirely accomplished in the Decorated style, Patrington was completed with a spire and east window in the Perpendicular style, following the disruption of building work caused by the pestilence. This Humberside corner of old Yorkshire contains a high proportion of the region's best Decorated work, as displayed in the nave of St Augustine, Hedon, a church located in another once-lively trading centre, the tower at Ottringham, and the innovative use of brick in the choir and transepts at Holy Trinity, Hull.

The fourteenth century witnessed the building or rebuilding of many fine churches in Yorkshire; these are the equivalents of the 'wool churches' of the Cotswolds and East Anglia. Here, however, the wealth embodied in the splendid buildings of the lower vales derived not from the sale of cloth, but largely from the sale of wool to Italian merchants. One of the reasons why it is unusual to find intact examples of churches in a single medieval style in Yorkshire concerns the eruption of building work in the Perpendicular manner during the last two centuries of the Middle Ages. This could involve a remodelling, with the insertion of

At Skelton, Yorkshire boasts perhaps the finest small Early English church in Britain; its thirteenth-century architecture proved an inspiration to the church builders of the nineteenth-century Gothic revival. Although legend holds that it was built from leftovers from York Minster, it was probably commissioned by the Minster Treasurer, Roger Haget, in about 1240, employing some of the masons who worked on the transepts at York

fashionable new window tracery or the addition of a west tower; a more complete rebuilding; or the provision of a church where none had previously existed. Several communities in the Dales gained their first church during the long currency of the Perpendicular style. From late medieval through to Stuart times, the so-called 'Tudor' version of the Perpendicular style was often adopted, with square-headed mullioned windows, as in the chapel of the church of St Gregory, Bedale. The 'churchwarden' style of modest buildings, echoing the simplicity of small twelfth-century churches, appeared later, particularly in Cleveland and areas just to the south. Ingleby Arncliffe, near Northallerton, is a good example, dating from 1821, and Ingleby Greenhow, near Stokesley, is another; the Norman church at Arncliffe, in

Langstrothdale, was restored in the churchwarden style at the end of the eighteenth century. In the English lowlands, most localities were served by a church which was already ancient by the time that the medieval period closed; in the more rugged parts of Yorkshire, however, the gradual process of providing new churches continued, with some townships acquiring a Methodist chapel before their church was built.

If a 'Yorkshire character' is apparent in the region's churches, this is partly a reflection of the use of local building materials. Few parts of this region were remote from sources of building stone, most could tap some adequate supply, and some areas had access to stone of the highest quality. In the west, the churches are usually of sandstone, the coarse, durable Millstone Grit or the less gritty and more friable sandstones obtainable from the Yoredales or the Coal Measures. Churches in the Vale of York – and splendid abbeys like Fountains and Byland – could exploit the superb building stones of the Magnesian Limestone belt which runs along its western edge, and stone from the renowned limestone quarries near Tadcaster was exported to build fine churches in the south-east of the region, at Beverley, Hull, Howden and Hedon. The North York Moors offered a variety of Jurassic building stones; only in the chalky Wolds and the eastern fringes did masons sometimes have to work with cobbles and shingle. While the churches of Yorkshire tend to be less ornate than some in the southern counties, they do tend to display good workmanship, a careful consideration of proportions, and a measure of restraint in matters of decoration. All the medieval styles were imported from the Continent via proving grounds in the South and Midlands; when assimilated into Yorkshire churches, however, these styles often reflected regional preferences. Influences from Somerset affected church building in the more prosperous south of Yorkshire at the end of the fourteenth century; although the Perpendicular towers of Somerset are unsurpassed, masons working in Yorkshire did modify the designs. They tended to retain the standard of having two belfry openings of two or three lights each, but kept embellishments to a minimum until reaching the parapet, where the regional enthusiasm for two tiers of arcading and crocketed pinnacles, tall at the corners, was unleashed. Good examples can be seen at Skirlaugh, near Hull, and Conisbrough, Fishlake and Tickhill, near Doncaster. Addison attributed the penetration of Somerset influences to two Yorkshiremen, Walter Skirlaw and Nicholas Bubwith, who had both become Bishops of Bath and Wells, and he added that: 'It would be too much to expect that even Yorkshiremen so distinguished as these two would get away with importing south country styles into Yorkshire unmodified. In fact, they didn't. In most of these churches modelled on Somerset styles, what were to become Yorkshire features broke through.'

The most distinctive group of churches are those in the 'Pennine Perpendicular' style, which are most numerous in the Craven area. These long and low churches date mostly from the last century of the Middle Ages and may reflect a late phase of wool-based affluence. They are generally characterised by low-pitched roofs – perhaps a modification demanded by the use of heavy roofing flags – straight-headed 'Tudor' windows, battlemented towers, and the absence of chancel arches. Examples include Bradfield, near Sheffield, Grinton, in Swaledale, Skipton, Kirkby Malham and Hubberholme. Bottomley was sceptical about the existence of a Yorkshire style of church building, but recognised the distinctive nature of the churches of the western uplands: 'After looking at a number of exteriors, the explorer might ask himself whether there is such a thing as a typically Yorkshire church. He could well answer in the negative but be aware that there are churches which do seem characteristic of particular parts of the county, for example, the long, low silhouette of the "Dales" church which is also often found in the West Riding.' He then proceededed to identify certain local idiosyncracies in church design: the corbelled-out parapets on several churches in and around Leeds; Decorated towers with eight rather than the usual four pinnacles in the East Riding; and the style of spires in the Malton area. More generally, Norman beak-head decoration was uniquely popular in Yorkshire, and the region's Romanesque doorways are highly regarded. There are around 150 Yorkshire entrance doorways surviving from the twelfth century and, of these, about 30 display figurative sculpture. Particularly fine examples include Healaugh, where the Day of Judgement is depicted, and Garton-on-the-Wolds, with St Michael and the dragon flanked by angels.

No church of any antiquity has survived without experiencing changes of one kind of another. The visitor with a practised eye can wander around a building and recognise fragments of Norman decoration and arches or tracery in the various Gothic styles. But no observer is skilful enough to recognise all the greater and lesser transformations – only a full-scale excavation can reveal these. At Wharram Percy, for example, excavation has revealed a dozen distinct building phases. Here and elsewhere in Yorkshire, rapid population growth in the twelfth and thirteenth centuries frequently meant that a church was too

*The elements of the Pennine Perpendicular style,
displayed in the church at Kirkby Malham*

small to accommodate the congregation and enlargement was necessary. In the thirteenth, fourteenth and fifteenth centuries, prosperity based on wool sales could fund another rebuilding and provide a tower for neighbours to envy. Changes of a different kind followed the Reformation, with the successive attempts to rid the buildings of 'Popish symbols'. A purge directed at rood-screens in the 1720s and 1730s deprived the region of almost all its screens; the fine example at Flamborough is a rare survivor. At Hubberholme, in remote Langstrothdale, the rood-loft, probably salvaged from Jervaulx Abbey at the time of the Dissolution, has survived, doubtless owing to the remoteness of the church, which caused it to be overlooked by authority.

The academic literature on churches is dominated by the approaches of the art historian; there is much less on the role of the church within the landscape and its function in relation to the local community. From the outset, churches were highly significant, central places within the rural context, often providing the nuclei around which settlement gelled. Examination of their tributary areas or parishes can provide insights into the former partitioning and organisation of territory; the existence of exceptionally large parishes, like those of Helmsley or of Grinton in Swaledale, has been noted. Excavations of churches, their interiors and surroundings, are, understandably, rather few, but where they have taken place, they have revealed an unexpected complexity in the evolution of quite modest churches and have provided information about the health, life expectancy and pathology of the members of the congregation who were buried in the churchyard. They may also show the ways in which the church was modified to take account of social, demographical and theological changes The apparently unremarkable church at Kellington, near Ferrybridge, was excavated in 1990–1 in the course of a scheme to dismantle and re-erect the tower when the church was threatened by coal-mining operations. It emerged that the church was preceded by a small Christian cemetery of the tenth and early eleventh century. The first church, erected about the time of the Norman conquest, was a timber building, built in the cemetery on a raft of limestone cobbles after the burials in its area had been exhumed. This timber building was swiftly replaced by a new church of stone, which was built between around 1080 and 1130. It was much larger than its predecessor, whose outlines were enclosed by its nave. In the twelfth century, mirroring the period of population growth, the tower was added; with the expansion of the nave soon

From the Reformation onwards, attacks on 'popish' symbols robbed Yorkshire churches of many of their medieval masterpieces. This alabaster carving of the Adoration of the Magi dates from the thirteenth century and was rediscovered in the nineteenth century; it is displayed in Burnsall church

afterwards, however, the tower had to be demolished and rebuilt, and changes in ritual and fashion resulted in the west entrance being replaced by entrances to the north and south. The tower was rebuilt again in the thirteenth century, when a north aisle was added and the chancel was lengthened. During the fourteenth century, changes in the local social hierarchy were revealed by the insertion of several high-status burials in the interior of the church, in the area of the north door. In the fifteenth century, the church gained a south porch, a clerestory and a chapel. The excavations made Kellington the most extensively excavated 'working' parish church. At Pontefract, excavations in the area of the Booths, to the east of the castle site, in 1985 revealed another cemetery which predated the building of an associated church; 197 burials were excavated, the earliest giving a radiocarbon date of AD 690, plus/minus 90. At Wharram Percy:

> About 600 burials were excavated, comprising about 1,000 individuals. There were several superimposed layers of graves, and many earlier burials had been disturbed by later ones. The graves were regularly laid out in rows, and the graveyard seems to have been used for at least four cycles of burial as the ground level had built up, especially on the north side. To the west the depth of soil was less. All this material is still being worked on, but it is hoped that it will produce important evidence for the physical anthropology, mortality, disease and nutrition of a medieval rural population to compare with evidence from urban populations in York.

At Ripley, in Nidderdale, the excavation of the churchyard at the former village site took place in 1991–2 after 70 medieval graves had been disturbed in pipeline-laying operations. A total of 128 burials of the twelfth to fourteenth centuries were excavated, the bones undergoing pathological analysis at Bradford University before reburial at the church of the present village. The remains showed a population comparable in stature to modern villagers, but suffering from arthritis and tooth abcesses.

Many churches today receive more visitors

Before the Reformation, medieval churches were decorated with colourful wall paintings. This representation of the Descent from the Cross dates from about 1250 and is part of a larger pageant of paintings on the walls of Easby church

than worshippers, so it is important to remember that the Church has exerted an enormous influence on Yorkshire society. Monastic communities played a vital role in worship before the Reformation. The conversion of the region to Christianity was largely the achievement of Aidan and his followers in the Celtic Church, even though Christianity had probably survived since Roman times in British cultural bastions like Elmet. The early churches were minsters, staffed by small bodies of monks who took Christianity to the surrounding countrysides. In due course, scores of the parish churches that had been provided by estate owners fell under the control of greater churches or monastic houses. After the Norman conquest, it was common for the lords to undertake the rebuilding of a parish church and then to grant it to a favoured religious house. By the close of the twelfth century, interests in about 200 Yorkshire churches had been conveyed to monastic houses. Popular attitudes towards the organised Church varied from place to place and from time to time. In many places, the monks' expropriation of common lands and resources was deeply resented, and the wealth and grandeur of some churchmen also affronted some of the principles of Christian conduct: in 1179 the Archdeacon of Richmond was travelling with a retinue of 97 horses, accompanied by 21 dogs and

3 hawks, although the convention of the Church limited his retinue to 7 horses. The lesser clergy could also be the targets for gossip; the prohibition of marriage was sometimes flouted, and a thirteenth-century Archbishop of York was the illegitimate son of one John le Romeyn, the parson at Hampsthwaite. In 1323 the parson of Tankersley was one of a band of thieves who broke into the house of Adam Cosyn at Flockton and stole goods to the value of twenty shillings. Even so, day-to-day life was firmly enmeshed with the affairs of the Church. The abbeys and greater churches heavily influenced economic life, and most peasant communities worked close to, if not side by side with, ecclesiastical farming enterprises. The Church was a mighty landowner and taxer, a powerful landlord and employer. A great church or abbey was likely to control the parish church; where no such church existed, worship would depend upon the services provided at or around a monastic chapel, and the provision of education was monopolised by the monks. The Pilgrimage of Grace in 1536 articulated a variety of fears and grievances. Some of its northern supporters genuinely valued the strong presence of the Church in northern affairs, while others preferred a known enemy to the anticipated intrusion of privileged southern lawyers and entrepreneurs. A strong strand of resentment against the effects of

the Reformation combined with a distrust of southern government and influences to fuel the less effective northern rebellion of 1569. Thereafter, localised pockets of Catholicism survived in Yorkshire, tending each to be focused on noble families which clung doggedly and dangerously to the old faith.

The Protestant faith took root most readily in

*Refined and spectacular
Perpendicular tracery at
St Mary's, Beverley*

Hubberholme church epitomises the problems of supporting Christian worship in the more remote corners of the Dales. The church was a chapel of ease of Arncliffe, four miles away as the crow flies and further via rough hill tracks, and corpses had to be taken to Arncliffe for burial. In the late fifteenth century, a corpse was swept from its bearers by the raging waters of the Wharfe and eight other bearers also perished in snowdrifts, so Hubberholme was granted the rights of burial. Thanks to its remoteness, Hubberholme retained its rood-loft – the edict of 1571, which required the destruction of rood-lofts in the diocese of York, apparently failed to reach this backwater

the more prosperous areas in the south of Yorkshire; islands of Catholicism survived in the North Riding, and there was a remarkable concentration of adherents in Nidderdale, in the old West Riding. Though led by prominent members of the local landowning aristocracy, like the Ingilby and Beckwith families, this partly reflected the conservatism of the backwater and the importance of the services provided by the former monastic communities. At the close of the Middle Ages, one could have followed the Nidd for a good 20 miles (32km) between the churches at Hampsthwaite and Middlesmoor and one would have encountered only one other parish church (at Pateley Bridge). The four monastic chapels in the dale greatly facilitated Christian worship, and the presence of chantry priests in many pre-Reformation churches gave the other priests a greater freedom to wander and preach in the out-posts of their vast parishes. After the restoration of Charles II, in 1660, numbers of Puritan clergy were expelled; these dissenters, like the Catholics before them, developed their congregations furtively, with Presbyterian chapels appearing after the Act of Toleration of 1689. Quakers also became active in the second half of the seven-teenth century, when their courage and obduracy initially made them the targets of vicious persecution. In the eighteenth century, the apparently complacent and moribund character of the established Church caused a craving for evangelical reform, with the Methodist movement inspired by John Wesley struggling at first to reform the Church from within and then proceeding to establish new centres for worship. Between 1797 and 1849, no less than five major Nonconformist movements broke away from the Wesleyan Methodists, each establishing its own local footholds in Yorkshire, while changes in Presbyterian organisation gave rise to Independent or Congregational churches. As the 'competing Nonconformist sects worked to secure congregations, so the Church of England joined in the struggle to provide new places of worship. One important consequence of this was that the new

and growing industrial communities and the scattered populations of the Dales and uplands gained a far more generous endowment of churches and chapels than had existed in Yorkshire in late Tudor and Elizabethan times. Notable examples include the Anglican revival church of St Peter (1841) and the church of St Aidan, built in the basilican style in 1894, both in Leeds.

Finally, there are the great churches from which control and influence radiated. York Minster is arguably the most beautiful cathedral in the whole of Britain. The building has experienced many phases of rebuilding and stands upon the ruins of the headquarters of the Roman

York Minster, arguably England's finest cathedral

legionary fortress. The Saxon minster here was burned during the uprising of 1069; there is now some doubt as to whether it was the direct successor to the hastily erected wooden church where King Edwin was baptised in 627. The new Norman minster was a cruciform church with an apsidal chancel, though a fire in 1137 destroyed this apse. Substantial additions were made later in the twelfth century, but the building that is seen today is the product of a sustained succession of works accomplished between 1230 and 1470. The last upstanding part of the Norman church disappeared in the late fourteenth century, with the demolition and rebuilding of the choir, although the crypt beneath the presbytery still survives largely intact from the second half of the twelfth century. The ill-fated south transept, recently restored, was built in the fourth decade of the thirteenth century and then the north transept was rebuilt. Towards the end of the century, operations centred on the rebuilding of the nave, where work continued until around 1340, culminating in the glazing of the superb west window, which is one of the great accomplishments of the Decorated style. The building was completed with the addition of the three Perpendicular towers between 1430 and 1470. Fortunately, the location close to the Ouse greatly facilitated the assembly of building materials. The stones were quarried at Thevesdale, Bramham, Stapleton and Huddleston and were carted to nearby river ports at Tadcaster, Wheldrake and Cawood respectively, where they were shipped by barge to wharves by the Ouse and finally brought to the minster site by sledge. This remarkable church was not conceived as an entity, but grew organically in the course of the medieval period. The quest for perfection continued for centuries; masons, carpenters and other members of an evolving army of craftsmen were almost continuously at work from the Norman to the Tudor ages.

In contrast to the ornate and rather French appearance of York Minster, Ripon Minster may seem a somewhat cliff-like and monolithic building. However, it may have a longer history, for entombed beneath the crossing there survives the crypt of Archbishop Wilfrid's church, dating back to around 670. Here pilgrims could view the assembled religious relics and so complete their journeys; Wilfred had brought back relics from his visits to Rome, and it is possible that his body was rested here after its transportation from Oundle, where he died in 710. Norman work of the late eleventh century survives in the adjoining apsed chapter house, but a wholesale rebuilding was launched a century later, employing the Transitional style. This had evolved into the Early English style by the time that the west end was reached. The collapse of the east end in 1288, followed by the partial collapse of the tower in 1458, necessitated rebuilding; the reconstruction of the nave early in the sixteenth century added to the somewhat incongruous appearance of the architecture, and the removal of the wooden spires which crowned the towers in the seventeenth century may not have improved the external appearance. Ripon Minster became a cathedral in 1836, regaining the status that it had enjoyed in the seventh century. The only church in England with a west front to rival, or even surpass, the splendour of York Minster is Beverley Minster. St John Beverley is said to have founded a monastery here in the eighth century, and in 935 it was refounded as a college of secular canons. A Norman church replaced a Saxon predecessor, but the collapse of its central tower in 1213 provided the catalyst for a great new building phase which spanned the eras of building in the Early English and Decorated styles. Building in the former style is beautifully displayed in the transepts; the nave is a magnificent essay in the latter; while the twin towers of the west front exemplify the Perpendicular ideal, completing the building and making it possible for one to appreciate the best that each Gothic style has to offer while walking from one end of the church to the other.

Perpendicular towers flank the spectacular west front of Beverley

REFERENCES

W. Addison, *Local Styles of the English Parish Church* (London: 1982).

M. W. Beresford and J. G. Hurst, *Wharram Percy* (London: 1990).

F. Bottomley, *Yorkshire Churches* (Dover: 1993).

S. A. J. Bradley, 'The Norman door of St Helen Stillingfleet and the legend of the holy rood tree', in P. V. Addyman and V. E. Black (eds), *Archaeological Papers from York Presented to M. W. Barley* (York: 1984), pp. 84–100.

J. E. Burton, 'Monasteries and parish churches in eleventh and twelfth-century Yorkshire', *Northern History*, 23 (1987), pp. 39–50.

P. E. H. Hair, 'The chapel in the English landscape', *The Local Historian*, 2l (1991), pp. 4–10.

J. G. Hurst, 'Wharram Percy: St Martin's Church', in P. Addyman and R. Morris (eds), *The Archaeological Study of Churches*, Council for British Archaeology Research Report 13 (London: 1976), pp. 36–9.

V. Leigh, *Outstanding Churches in the Yorkshire Dales* (Settle: 1983).

R. Morris, *Cathedrals and Abbeys of England and Wales* (London: 1979).

H. Mytum, 'Kellington Church', *Current Archaeology*, 133 (1993), pp. 15–17.

A. D. Phillips, 'Excavations at York Minster, 1967–73' *The Friends of York Minster Fourth Annual Report* (1975).

P. A. Rahtz, 'The archaeology of the churchyard', in P. Addyman and R. Morris (eds), *The Archaeological Study of Churches*, CBA Research Report 13 (1976), pp. 41–5.

G. Randall, *The English Parish Church* (London: 1982).

W. Rodwell, *The Archaeology of the English Church* (London: 1981).

W. Rodwell and J. Bentley, *Our Christian Heritage* (London: 1984).

A. Wilmott, 'Pontefract' *Current Archaeology*, 106 (1987), pp. 340–4.

R. Wood, 'The Romanesque doorways of Yorkshire, with special reference to that of St Mary's Church, Riccall', *Yorkshire Archaeological Journal*, 66 (1994,) pp. 21–50.

CHAPTER FIFTEEN

Domestic Vernacular Architecture

DWELLINGS SHOULD ALWAYS INTEREST the landscape historian. Not only are they the building blocks of which settlements are composed, but they also mirror the conditions of class, economics, transport and technology as they existed at the time that they were built. Of no less importance is the fact that, in the lifetime of a dwelling, alterations and additions would be made to accommodate changing ideas about status and lifestyles; the diffusion of innovations and cultural influences was also mirrored in a region's dwellings.

Iron Age dwellings have been discovered at a number of places in Yorkshire; in 1973–81, a number of these roundhouses were excavated at a site near Roxby in the north of the North York Moors, just inland from Staithes. The dwellings that were excavated in most detail showed some differences in construction; they were circular, with wall diameters of 7–8 metres. The roofs, which must have been conical and thatched, were supported by internal rings of posts; drips from the thatch were caught by a circular gulley, while the house walls could be composed either of stakes or of low stone walls. The use of stone would greatly prolong the life of a house; one example here, with stone walls and strong posts, may have been built in Roman times and have endured into the sixth century, though it might have degenerated into a timber shelter towards the end of its useful life. Its 50 square metres of floor space could have housed a family group of about five members. The paucity of the finds inside – 140 pottery sherds and the base of a beehive quern – were thought to suggest that its occupants led subsistence lifestyles, which, along with its situation in an abandoned arable field, led to its interpretation as a herdsman's family

home that may only have been occupied seasonally. Three other dwellings seem to have formed a cluster within the settlement and might have accommodated the households of an extended family. Two produced evidence of occupation and industrial work, like smithing, but one house seemed only to be associated with the activities of women, such as milling, so there might have been some specialisation in the use of buildings. Arable land lay to the north of the group of houses, with grazing land to the south; the excavators considered that:

> The work at Roxby has shown that the Iron Age people in north Yorkshire were capable of a full mixed economy of cultivation, pastoralism, metallurgy and other skills. Only the relative crudeness of their pottery and, it might be argued, the paucity of artefacts found in the houses and lack of a money economy, differentiate them from their counterparts in southern England. Moreover they achieved this on tough boulder clay fields and relatively infertile moorland pasture.

Spratt added that: 'If they have managed so well on the marginal land at Roxby, one wonders how rich were the settlements on the more fertile ground.'

In the course of the Roman period, dwellings of the roundhouse type, which had been built in Britain since at least late Neolithic times, were gradually replaced by rectangular structures, which were often subdivided to provide two or more chambers. Anglo-Saxon domestic architecture involved buildings with walls of posts and wattle and daub or of split oak logs, and two main types of building are recognised, the hall and the

The popular conception of the 'typical' Yorkshire home. This fairly early example of a gritstone laithe house is dated 1673

Grubenhaus, which has variously been interpreted as accomodation for serfs, a weaving shed used mainly by women, and a grain-storage shed. Invaluable evidence of the Dark Age development of domestic buildings in Yorkshire resulted from the excavations at Coppergate, York. Early in the tenth century, rectangular buildings with walls of wattle tightly woven around posts were erected, but later, during the Scandinavian occupation, quite different techniques were employed. Hall writes that:

> On all four tenements the wattle-walled houses-cum-workshops were rebuilt in a completely new style in which a semi-basement as much as 1.5m deep replaced a floor at ground level, and at least the lower 1.8m of the walls was fashioned from substantial oak planks and posts. Although these semi-basements may seem a peculiar type of building, the nature of the floor deposits indicates that they were lived in, and were not just cellars for storage.

Two ranks of these buildings were recognised on two of the tenements excavated, and it was surmised that the front, street-side rank served as living quarters, perhaps equipped with a stall for selling goods to passers-by, while the rear rank of the buildings functioned as workshops. The configuration of the excavation site meant that, only in two cases, were the full lengths of buildings exposed:

> These measured 7.5–7.6m x 3.8–4m, but it does not follow that the street frontage buildings were identical in length, and as main living quarters they may have been rather longer. This type of building seems to have been first constructed at Coppergate in the period *c.* 970–80, when they appeared on several different tenements, again suggesting that, as with the post and withy buildings at the start of the century, there was a simultaneous reconstruction across the four excavated tenements.

The evidence for early medieval building methods is uncertain; timber dwellings were built at Wharram Percy in the twelfth century, but the villagers lived in small one- or two-roomed dwellings in the fourteenth century. In some cases, the walls were built of chalk blocks; in others, the chalk-block footings may have carried a wall of wattle and daub. The chalk was easily obtained, with each family excavating materials from their own 'backyard'. Roofs would have been of some form of thatching, carried on ash pole- rafters. The outstanding feature of these homes – apart from their poky natures – was their flimsy construction, so there may have been a need to rebuild about every generation or so, possibly resulting from a lack of professional expertise in roof construction. Thus, with the successive rebuildings, the house would gradually migrate around its fixed house plot. The families here kept their homes as clean and tidy as possible, and their frequent sweeping of the floors eroded the packed earth and undermined the walls. Windows, one can presume, were small, unglazed and fitted with shutters, but a rather surprising sophistication was the general use of door hinges, locks and keys. In the fifteenth century, some peasant families at Wharram replaced their boxy little dwellings with longhouses, around 27 metres in length. Dwellings of this kind had a much more extended history than this, and they continued to evolve in Yorkshire through to the eighteenth century. They were long, low buildings, often with just one living-room, with the remainder of the house consisting of a byre. The cause of the shift to longhouses at Wharram is uncertain. Perhaps the keeping of animals became more widespread, or perhaps the worsening of the climate made it necessary for animals to be

The Bowes Morrell house on Walmgate, York, dates back to the late fourteenth century; the house plan is unusual for the period

wintered indoors. In any event, the close proximity of sheep, horses, or oxen would have provided an effective, if rather insanitary heating system. Beresford and Hurst write that:

> At the time of desertion, about 1500 . . . each toft contained a single long house or a simpler building. Excavation has shown that long-houses go back to at least the thirteenth century and that simple two-roomed houses, and long-houses in which the farming activities took place under the same roof as the living accommodation, were in use at the same time. It is suggested that the choice depended simply on the prosperity of the peasant and whether or not he had a number of animals.

The use of stone walling in the houses at Wharram – a village which could only afford to employ the soft local chalk – would, during the thirteenth century, have superseded an earlier tradition of employing poor timber frames. This seems to have been almost a nationwide transition; the next switch in peasant building, which affected most of Yorkshire, was a return to (an improved type of) timber framing. This was based on the 'cruck framing' method, which probably dates back to at least the thirteenth century in the southern counties of England. At each end of a bay, support was provided by a pair of heavy wooden blades, joined at the top and splayed out at the bottom to form an A-shaped frame. A house of just one bay would employ only two of these frames, but longer houses would have two or more such bays. The number of cruck or 'crock' frames provided a convenient shorthand method for describing a dwelling. Thus, a survey of houses near Ripley in 1635 records that William Raynard had a 'house with barn in the west end of 3 pairs of crocks, parlour chamber and oven house in good repair £1 12s. 0d. He holds 3 acres, 2 rods, 22 perches of land worth £3 6s. 8d. per annum.' In due course, practices changed again: cruck-framed dwellings and barns were encased in stone, and newer ones were built solely of stone. The antiquity of the cruck-framing method in Yorkshire is uncertain, but it had taken root well before the end of the Middle Ages. A lease of 1432 describes a house of eight crucks (that is,

three bays), roofed in stone slates, in Ovenden township; in 1495 the Abbot of Whitby had oaks felled for cruck blades to repair his house at Goatland; and, in the sixteenth century, cruck-framing seems to have been the norm in most parts of the Dales, the North York Moors and much of the south of the region. In South Yorkshire, with 158 known cruck buildings, including 94 barns, Ryder (1980) concluded that the oldest known cruck-framed buildings dated from the fifteenth century, and the youngest to the late seventeenth or even eighteenth century. He writes that three main concentrations of crucks are apparent in England:

> The second of the major cruck concentrations, the Pennine fringe area, runs through the western part of South Yorkshire, which provides one of the densest concentrations of all. All but a dozen of the 150 known cruck buildings in the County are confined to its western third, the Pennine moors and valleys west of a line from Sheffield to Barnsley . . . The reasons for this 'astonishing sharp division on the eastern boundary between cruck and non-cruck areas' (Smith 1975 p. 3) have long been debated. Smith points out that this boundary does not correspond with any known political or cultural frontier, but is unable to offer any reason for such a sharp delineation in such a small area.

In the old West Riding, however, the evolution may have been different from that outlined above; though excavations and very old dwellings are fewer here, the documentary record has been carefully explored. The medieval rough-stone phase of building may not have occurred here, since mentions of carpenters are much more numerous than records of masons. Timber framing seems to have continued right through the Middle Ages, and ordinary dwellings appear to have been flimsy and, possibly, prefabricated: there are records of homes being carried away as well as others being knocked over. Some higher- quality late medieval houses survive, but most show the fashion for encasing timber frames in stone during the seventeenth century.

The homes of the dominant elements of medieval society were, obviously, quite different from those of the rural masses. Because of the natural limits governing the size of timber available to construct cruck blades, this method of framing, whatever its antiquity, would normally only permit the building of modestly sized homes of a long but narrow configuration. To build on a more imposing scale, one needed either to work in stone or to employ the box-framing or post and truss method, where the vertical support was provided by stout upright oak posts which could carry the superstructure, be stabilised by bracing timbers or be augmented by upright timber studs.

For a durable building to result, the posts would need protection against rotting; a study by Le Patourel of the sites of moated medieval houses suggests why early medieval survivals are few. At first, the posts were set in holes in the ground and were very prone to rot at ground level. In the thirteenth century, the post base was placed on a flat stone, but was still set below ground level, though the post hole might contain a packing of stones. Then the post was placed on a flat stone laid on the surface of the ground; next, in the fourteenth and fifteenth centuries, posts were jointed into a

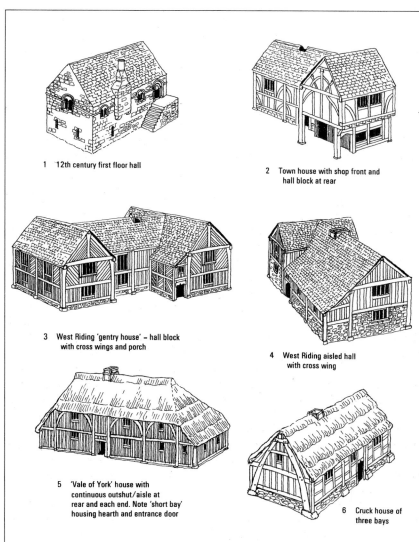

1 12th century first floor hall

2 Town house with shop front and hall block at rear

3 West Riding 'gentry house' – hall block with cross wings and porch

4 West Riding aisled hall with cross wing

5 'Vale of York' house with continuous outshut/aisle at rear and each end. Note 'short bay' housing hearth and entrance door

6 Cruck house of three bays

Some of the characteristic medieval house types of Yorkshire.
(Peter F. Ryder, reproduced from *Medieval Buildings of Yorkshire*, Moorland Publishing Co. Ltd., 1982)

timber sill beam, which was carried on a low stone wall base, set in a shallow trench. It was in these ways that more permanent, rot-proof structures evolved.

If an early medieval building was constructed of stone rather than built around earth-fast posts, its prospects for survival were much better. A rare and very important example is Burton Agnes Old Hall, near Bridlington. This is a Norman first-floor hall and one of the best survivors in England of this distinctive kind of home; most others are to be found in towns – notably Lincoln and Bury St Edmunds – where they were occupied by wealthy members of the Jewish commercial community. The old hall was probably built around 1175 by

Roger de Stuteville, and is now most notable for its well-preserved undercroft, with cylindrical stone columns and quadripartite vaulting. This is slightly reminiscent, though on a much smaller scale, of the famous vaulted cellarium at Fountains Abbey. The undercroft would have been used for storage, with domestic life and the affairs of the estate being conducted above, at first-floor level, in the hall, which was reached by an external stone staircase. One end of the hall would have been partitioned off to form the private apartment or solar, used by the Norman lord and his family. Another first-floor hall, smaller and less intact, survived at Hooton Levitt, near Maltby. Both these buildings endured because they were able to serve as outbuildings when new manor houses were built nearby. It would be very difficult to discover how common such stone-built first-floor halls were in Norman Yorkshire, although the long continuation of the tradition

Burton Agnes Old Hall has a superb vaulted undercroft dating back to the late twelfth century

into the fourteenth century is apparent at Markenfield Hall, where the main affairs of the occupants were also conducted at first floor level. Another example is Scolland's Hall at Richmond Castle, probably built by Alan, Earl of Brittany, around 1071–89, where the solar survives.

In the centuries that followed, stone was still mainly reserved for churches and castles; those local aristocrats of less than the castle-owning rank tended to occupy timber-framed dwellings. In a few cases, the transition from timber to stone can be recognised and dated. In the early 1990s, a rescue excavation was undertaken at a medieval moated site at Rawcliffe, just north-west of York. The excavations revealed a moated site of the eleventh and twelfth centuries with a timber-built hall range; just over 100 metres away was found the stone footings of its replacement, a manor house of the thirteenth to sixteenth century, with an aisled hall. The casual traveller might think that, in comparison to regions like East Anglia and the West Midlands, Yorkshire was bereft of timber-framed houses. In fact, these houses exist in some numbers; were the region less well provided with cheap resources of durable stone, then these buildings would be apparent. However, the later fashion for encasing dwellings in stone shells which remodelled and updated their exteriors has led to the concealment of a considerable legacy. This concealment has made it difficult to compile a detailed chronology of building techniques for the region, although it does appear that Yorkshire carpenters persevered with methods which had become archaic in the 'progressive' South.

In general, the evolution of the house plan in Yorkshire seems to have been similar to other parts of England. Houses would be dominated by their halls – large public rooms, open to their rafters – which had a solar, with service or storage rooms beneath, at one end and, perhaps, kitchen and service accommodation at the other. As the medieval period progressed, such houses would tend to sprout cross-wings, giving extra domestic and service accommodation at one or both ends of the hall block. The hall remained the predominant component of the house throughout the Middle Ages in Yorkshire. Lees Hall, at Thornhill Lees, near Bradford, is a good example of a late fifteenth-century house, with hall and cross-wings,

in the solid timber framing of Yorkshire, as is Shibden Hall, Halifax.

Stone building was used for some of the greater medieval houses, as in the late thirteenth-century first-floor hall of Grassington Hall, the imposing surviving wing of Hipswell Hall, near Richmond, built in about 1500, and the magnificent Fountains Hall, built after the Dissolution from stone plundered from the deserted abbey. Lawkland Hall, in Craven, has a longer history; the tower and south front date back to the reign of Henry VII, though most of the sandstone house is of Elizabethan vintage. Where it was readily available, stone was used in town houses of moderate status from the end of the Middle Ages onwards. Richmond is acclaimed as one of England's finest stone-built towns; in the seventeenth century, the traveller Celia Fiennes remarked on the stone houses she had seen there: 'the streets are like rocks themselves'. One of the finest assemblages of timber-framed town houses in England can be seen in York, the dwellings dating from the fourteenth to the early seventeenth centuries. The restricted nature of the building sites probably combined with fashion and a sense of display to encourage the use of jettied or projecting upper storeys, with houses being built three storeys tall during the last three centuries of the Middle Ages, as the examples on Stonegate show. Two-storeyed jetty houses which are thought to date back to 1316 comprise Lady Row on Goodramgate. One of the finest buildings, both externally and within, is the Merchant Adventurers' Hall on Fossgate, with a brick and stone undercroft carrying a fine timber-framed superstructure of the fourteenth century. Although the houses on Lady Row were built originally for journeymen and labourers, most surviving medieval town houses were the abodes of members of the equivalent of the commercial middle class: artisans, merchants and shopkeepers. Meanwhile, the lower rural classes had to wait rather longer for the day when they could aspire to durable, serviceable dwellings. During the fifteenth century, the more prosperous members of the emergent yeoman classes began to acquire workmanlike and durable dwellings; like some of the greater houses of East Anglia, these are a legacy of the buoyant woollen industry. A number

Jettied houses in York's famous Shambles. The street is so narrow and the jetties so pronounced that neighbours could virtually shake hands across the street. In cramped urban situations, the jettied upper storeys allowed a useful increase in floor space, though, in days of rudimentary sanitation, pedestrianism must have been a thankless pursuit

of surviving examples are found in the Halifax region. They were built to the established tradition of the aisled hall, long exploited both in houses of various social classes and in a common form of barn construction, with the side-aisle or aisles being defined by a row of posts, comparable to the piers in a church. The medieval practice of centring public and domestic life on the hall and of managing with a minimum of more specialised rooms persisted, so the aisle may have served as sleeping accommodation In Pontefract, there is a pub and restaurant called the Counting House

which, until a recent restoration, was a derelict outbuilding: a survey by the local archaeological society showed that it probably dated from the decades around 1600. It is built robustly of close studding, with the timber framing set upon the stone walls of the lower storey. Like many old buildings, particularly those in towns, it has experienced a series of uses, perhaps being built for storage by wool merchants who lived in a house in front, then being used for malting, and, in the nineteenth century, being sectioned to provide two cottages for old people. A number of regional preferences were becoming apparent amongst the dwellings of Yorkshire, with king-post roofs and heavy roofing timbers being favoured in the Pennines region, and collar-rafter roof construction being popular in the lowlands from the fifteenth century. In both cases, the framing in yeoman or grander houses tended (like church masonry) to be sound, solid, but with restrained decoration; both styles also favoured close studding, with a virtual forest of oak being displayed in the closely spaced uprights or studs of the frame, while fan-like patterns of struts were often used in the gable ends of Pennine houses. Several of these preferences were embodied in the Counting House.

Timber-framing devoured enormous quantities of purpose-grown trees, so perhaps a shortage of timber influenced the gradual switch to stone in the post-medieval farmsteads of the uplands. Longhouses, or 'coits', have a long but indeterminate history, as noted above. Their adoption at Wharram in the fifteenth century has been described; in some other places, however, they existed by the start of the fourteenth century. The records of the manor of Wakefield for this period reveal a spate of thefts of animals from houses, implying buildings of the longhouse type and showing that the common people variously shared their homes with sheep, oxen, horses, goats and cockerels. The longhouse could be built of timber, stone or both materials; an important component of the design was the cross-passage, which ran from the door of the house, between the living-room and the byre. A few stone long-houses survive in the north-west of the region, normally as ruins or converted into outbuildings, but the vast majority of the remaining farm-steads of the Dales are stone laithe houses, taking their name from an Old Norse word for a barn. Whether the laithe house evolved from the long-house is uncertain, but here the adjoining house and byre (or barn) had separate entrances. The oldest surviving laithe houses date from around

Merchant Adventurers' Hall, Fossgate, York. The brick and stone undercroft carries a timber superstructure displaying both arched braces and close studying, dating – apart from the roof and windows – to 1357–68

1650 in the West Riding and a little earlier in the northern dales; these buildings were extremely popular in the eighteenth century.

In the affluent South of England, the 'Great Rebuilding' had provided durable, workmanlike houses for many members of village and farming society during the Tudor period. However, the rising standards and expectations of living penetrated the North only gradually. To make matters more complicated, these improvements affected different classes at different times, as Ryder (1987) explains:

Although there are a wide number of types and styles of timber-framed building within South Yorkshire, some general patterns are perhaps observable. In the medieval period proper only the gentry could afford to build substantial timber houses, and few of these survive today ... In the second half of the fifteenth century, with increasing prosperity, the whole picture changes; many of the lesser gentry rebuild their residences, and yeoman farmers find themselves able to build houses on almost the same scale ... By the early sixteenth century, if not before, a third group were building in the west of the county, using the cruck frame, perhaps in continuance of the form of construction of the much-discussed medieval peasant house of which so few examples survive ... Despite the persistence of what might almost be termed a sub-tradition of hybrid timber-framed/stone-walled buildings in the Pennine West, framed construction seems to have continued in general use throughout South Yorkshire until the second quarter of the seventeenth century and the 'Great Rebuilding'.

Although there was a spate of building in the late seventeenth century, in parts of the Dales it was only in the eighteenth century that small farmers began to obtain dwellings of the laithe-house type, which could be considered habitable today. The first laithe houses were built to functional designs, but the later ones reflected the long-established fashion in upper-class housing for a symmetrical façade. Symmetry was then incorporated into the domestic component with a centrally placed doorway and balanced window openings, but it could not easily be extended to the adjoining barn. Symmetry was often retained in the nineteenth-century farmhouses, but the owners now lived aloof from their stock, and house and barn were built as separate structures. Meanwhile, the convenience of the double-depth or double-pile plan was accepted, and square houses which were two rooms wide were favoured. They certainly offered a more varied and convenient suite of rooms, but seldom seemed to blend into the countryside as comfortably as the rugged laithe houses.

Although the full extent of Yorkshire's heritage of medieval and Elizabethan buildings is masked by the convention of encasing timber framing in stone shells, far fewer old buildings have survived in the North of England as compared to the South. Because of the relative poverty of rural Yorkshire, particularly the uplands, the great rebuilding was delayed for centuries; insubstantial and perishable dwellings were still being provided for the poorer classes in the seventeenth and even in the eighteenth centuries. These points become evident from the results of a survey of vernacular buildings in North Yorkshire and Cleveland – regions in which virtually no such work had been done before. Hutton reported that:

The great majority of the houses we have recorded date from the 17th and 18th centuries. Only 91 out of 665 have features going back much earlier than 1600. If we consider a medieval house-type quite common in southern England, the hall and crosswing house, we can find only five examples with open halls, and eleven others with storeyed halls, some of which may possibly replace earlier open halls. But among those five open-hall houses, which are all manor houses, are the only two base-cruck halls known north of the Trent although about 125 have been recorded south of it.

The vernacular houses of Yorkshire were built with local materials The stone tradition gave rise to a varied legacy of buildings, each locality having its own materials, tones and textures. Among the numerous local choices were limestone rubble in Teesdale; walls and roofs in

St William's College, York, stands close to the Minster and was founded in 1461 for chantry priests. It has experienced a number of alterations and a nineteenth-century restoration

Yoredales sandstone in Swaledale; gritty sandstone in Nidderdale; softer sandstone in parts of the old West Riding; and fine, pale limestone in places, like Knaresborough, on or beside the Magnesian limstone belt. During the nineteenth century, the extension of the railway network made cheap Welsh or Cumbrian slate available, and the traditional roofing materials have disappeared from many roofs. Few thatched houses and barns can be seen in the Dales today, although they were still very common a century ago. During the Middle Ages, bracken and straw were both popular thatching materials, and a small number of buildings roofed in turf or heather can still be seen in the more remote places, mainly as ruins. Before the arrival of imported slates, 'thakstones' were quarried in various localities which possessed tough but fissile sandstones; where they survive, such roofs add greatly to the charm of the home beneath.

Houses using heavy thakstones from the outset tend to have roofs of a much shallower pitch than those which used thatch. In the Dales, many heather-roofed houses were re-roofed in thakstones in the first half of the nineteenth century. To lighten the load, the eaves were raised to create a shallower roof pitch, and one byproduct of such conversion was often the creation of an upper storey with proper bedchambers. The thakstone tradition had a long history, and there are numerous records of the quarrying of thakstones in the early fourteenth century. Wooden shingles were also used at this time; in 1315 Agnes de Newebyggyng was fined at Wakefield

Yorkshire embraces a number of vernacular traditions; the gritstone of the Dales contrasts with the limestone and pantiles tradition of the Vale of York, as represented by these houses in Helmsley

Manor court for hitting Margery the Wright over the head with one. Though rich in stone, the West Riding tended to use timber frames for smaller houses throughout the Middle Ages. Here, and in the Dales and North York Moors, structural walls of stone gradually replaced timber framing in the homes of members of the farming community during the seventeenth and eighteenth centuries. People who lived in the Vale of York and the Wolds, however, were often too far from suitable supplies of building stone, and constructional timber was often hard to obtain, too; the result was a legacy of houses in locally made brick, roofed in local clay tiles or pantiles. This facet of the Yorkshire heritage of buildings is often overlooked, although it is apparent in scores of villages with attractive russet dwellings of the eighteenth century. In fact, Yorkshire has a remarkably long tradition of building in brick. In 1293 Edward I purchased the hamlet of Wyke and renamed it 'King's Town upon Hull'; a decade later, Hull supported a municipal brickyard, becoming possibly the first English town to contain a substantial number of brick houses.

REFERENCES · ·

N. W. Alcock, *A Catalogue of Cruck Buildings* (London: 1973).

M. W. Beresford and J. G. Hurst, 'Wharram Percy: a case study in microtopography', in P. H. Sawyer (ed.), *English Medieval Settlement* (London: 1979).

D. W. Clarke, 'Pennine aisled barns', *Vernacular Architecture*, 4 (1973), pp. 25–6.

M. L. Faull and S. A. Moorhouse (eds), *West Yorkshire: An Archaeological Survey to AD 1500* (Wakefield: 1981).

J. A. Gilks, 'Boothtown Hall: a fifteenth century house in the parish of Halifax', *Yorkshire Archaeological Journal*, 46 (1974), pp. 53–81.

R. Hall, *The Excavations at York: The Viking Dig* (London:1984).

B. J. D. Harrison and B. Hutton, *Vernacular Houses in North Yorkshire and Cleveland* (Edinburgh: 1984).

R. H. Hayes and J. G. Rutter, *Cruck-framed Buildings in Rydale and Eskdale*, Scarborough and District Archaeological Society Research Report 8 (1972).

B. Hutton, 'Timber framed houses in the Vale of York', *Medieval Archaeology*, 17 (1973), pp. 87–99.

B. Hutton, 'North Yorkshire vernacular houses', *Current Archaeology*, 74 (1980), pp. 92–5.

R. Inman, D. R. Brown, R. E. Goddard and D. A. Spratt, 'Roxby Iron Age settlement and the Iron Age in north-east Yorkshire', *Proceedings of the Prehistoric Society*, 51 (1985), pp. 181–214.

H. E. J. Le Patourel, *Moated Sites of Yorkshire*, Society for Medieval Archaeology Monograph 5 (1975).

H. E. J. Le Patourel and B. K. Roberts, 'The significance of moated sites', in F. A. Aberg (ed.), *Medieval Moated Sites*, Council for British Archaeology Research Report 17 (London: 1978), pp. 46–55.

J. McNaught, 'Swales Yard, Pontefract', *Current Archaeology*, 141 (1994/5), pp. 350–1.

R. Morgan, 'Dendrochronological dating of a Yorkshire timber building', *Vernacular Architecture*, 8 (1977), pp. 809–14.

A. Raistrick, *Buildings in the Yorkshire Dales* (Clapham: 1976).

RCHME, *York: Historic Buildings in the Central Area* (London: 1981)

P. F. Ryder, *Timber Framed Buildings in South Yorkshire*, South Yorkshire County Archaeology Monograph 1 (Barnsley: c. 1980).

P. F. Ryder, *Medieval Buildings of Yorkshire* (Ashbourne: 1982).

P. F. Ryder, 'Five South Yorkshire timber-framed houses', *Yorkshire Archaeological Journal*, 59 (1987), pp. 51–82.

J. T. Smith, 'Cruck distributions: an interpretation of some recent maps', *Vernacular Architecture*, 6 (1975).

D. A. Spratt, 'Roxby', *Current Archaeology*, 105 (1987), pp. 308–11.

York Archaeological Trust, 'Rawcliffe', *Current Archaeology*, 140 (1994), p. 311.

CHAPTER SIXTEEN

Enclosure and the
Post-Medieval Countryside

IT TOOK SEVERAL CENTURIES for communal systems of open-field farming to be established in Yorkshire; such arrangements probably continued to be created in the twelfth and even thirteenth centuries. In some parishes, the basic elements of the system were still functioning at the start of the nineteenth century, but in other places members of the community had begun to unpick the complex patterns of communal farming almost as soon as they were introduced. In a great many parts of the Dales open fields had disappeared by the sixteenth or seventeenth century, enclosed pasture was widespread, though some fields were still shared by groups of tenants, and vast tracts of upland remained as common grazings. During the same period, enclosure by agreement amongst the tenants parcelled up most of the ploughland on the fringes of the North York Moors. For all the harsh injustices of the feudal system, it did provide the accredited members of a farming community with the opportunities for survival, but since some members of these communities were more enterprising or devious than others, village society often came to contain families who were busy creating their own small farming empires by extracting their lands from the semi-communal pool of resources. Where a village or township contained several such families, a swapping-around of strips and fields might allow a number of compact properties or tenancies to result from agreements and purchase. Frequently, however, success for one or two families was achieved at the expense of numerous less prosperous neighbours, who had cause to feel aggrieved. Such actions gave rise to a multitude of disputes, particularly when villagers claimed

that the common rights of pasture on newly enclosed land persisted during the months following the harvesting of the crop. In 1316, for example, Robert Carpenter fenced off his share or dole in a meadow on Wakefield Manor, depriving his neighbours of autumn grazing. The conflict between public and private countryside is a timeless theme, but one which has been felt most keenly at times of enclosure: in 1615, the 20 freeholders of Langfield complained that the best half of the waste there had been enclosed by the lord, depriving them of their rights to pasture and peat-cutting. In still other cases, as in parts of Swaledale in the sixteenth and seventeenth centuries, tenants engaged in enclosing intakes from the common for cow pastures, and it was their lords who felt cheated, since they were not obtaining any extra rent from the piecemeal new enclosures. On the North York Moors, 'orthostatic' walls (with their bases made of large, erect boulders of Jurassic sandstone which are bedded into the earth) can be found in both upland and valley settings. Although they appear prehistoric and resemble walls built in Derbyshire and other parts of Yorkshire during the medieval period, they are frequently associated with the enclosing of intakes during the period 1550–1750. In some places, they cross older field banks, and researches by Spratt show that some examples were still being built after 1782. The very varied fortunes of open-field farming can be read in the patterns of surviving walls and hedgerows. In scores of places, one can recognise fields with curving boundaries which are clearly enclosed parcels of strips, while close by there may be the straight, criss-crossing boundary patterns produced by the parliamentary

Field patterns in Upper Nidderdale. The higher ground displays the rectangular geometry of parliamentary enclosure, although the irregular fields below tell of an earlier phase of enclosure

enclosure of open fields and commons in the period of about 1750–1850.

The event that brought the medieval period to its close and which had profound repercussions on the rural landscape was the Dissolution of the monasteries. The monastic lands passed to the Crown, but rocketing inflation and the costs of wars against Scotland and France necessitated a disposal of monastic estates. Some lands were bought by speculators and self-seeking officials, others were obtained directly by local landowners seeking to enlarge their estates. In general, after the larger speculators had taken their profits, the old monastic estates tended eventually to become fragmented, passing into the local ownership of a mixture of former tenants and lesser gentry. The inflation that had argued for the sale of the monastic lands could have unfortunate consequences for landlords. Rent paid in kind or service had long since been replaced by cash rents (although, even in the reign of Henry VIII, tenants on some northern estates were required to serve in campaigns on the Scottish borders). Customary tenants, paying fixed rents and passing on their holdings to heirs (or even selling them), were the successors of the medieval peasants, but in inflationary times the landlords struggled to replace the fixed rents, declining rapidly in real

value, with leaseholds. Often, they reacted to the difficulties posed by customary rents by trying to recoup some extra income by exacting archaic feudal fines. Gradually, however, the medieval countryside of common fields worked by feudal peasants was being superseded by one of enclosed fields worked by lease-holding or copy-holding tenants.

Landowners were obviously interested in making profits and in enjoying the comforts which money could buy – but they were also keen on displaying the status of their dynasty, impressing their equals and entertaining their betters. The inevitable obsolescence of the private castle was paralleled by the rise of the stately home; accompanying this rise was the desire to provide a deserted, tranquil and carefully manicured setting which would present the house to its finest advantage. In some cases, as at Ripley, the fashionably landscaped park could be accommodated in an old deer park; at nearby Nidd, however, the village was eventually removed to

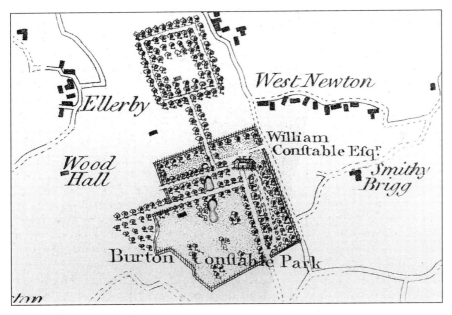

The age of the landscaped park at Burton Constable, as mapped in about 1770

produce the splendours of isolation which were now demanded, disappearing in three stages in the course of the nineteenth century. Even the lane that had been the former High Street was excavated to become a holloway sufficiently deep to prevent foot travellers from glimpsing the bordering park. The most famous case of emparking in Yorkshire involved Castle Howard: the old Howard seat burned down in 1693, and the construction of the mansion and its park and lake took place at the expense of the unfortunate villagers of Hinderskelfe. Villagers at Harewood were more fortunate; in the 1760s, while Harewood House was being built almost on top of their old homes, they were able to occupy the showy new stone terraced houses provided by the roadside, just outside the park gates. At Sledmere, in the Wolds, the landscape was drastically remodelled by the Sykes family, merchants of Leeds and Hull, who acquired estates there in 1718. In 1748 Richard Sykes inherited the estate and began to build a mansion beside the parish church, near the site of the previous house; after 1772, the village was removed to create the associated great park, its central section having already been removed to clear a prospect looking south-eastwards from the hall. The villagers were rehoused in a new settlement, built outside the park on the Malton–Driffield road. The old village lay on both sides of a road running to York and lying to the east of the hall and church, with other dwellings on the north side of a lane linking this road to the hall. In 1776 Sir Christopher Sykes presented a bill to enclose Sledmere to the House of Commons. The owner of the neighbouring estate objected, complaining that the road to York, Malton and Driffield had been blocked and his tenants obstructed from reaching Sledmere church, but this did not prevent the diversion of the York road and the enlargement of the park.

Fashions in the landscaping of parkland changed, with the geometrical formality of the earlier designs generally being superseded by the more naturalistic curves, lawns, lakes, spinneys and tree stipple favoured by advocates of the Romantic and Picturesque movements. Consciously or otherwise, these arbiters of taste were frequently re-creating the appearance of the old medieval deer parks, most of which had lapsed before the close of the Middle Ages. In more recent times, the preservation of unproductive parkland has become an impossible luxury and the region contains numerous patches of disemparked land, with park walls and some tree groupings standing, incongruously, amongst working countryside. The titivation of the countryside in the eighteenth and nineteenth centuries was not confined to the landscaped parks, but also involved the planting of small woods, screens and shelter belts, with Scots pine, larch and beech being generally regarded as the most suitable and tasteful species. Such woods could harbour game, provide a little timber, shelter fields, or obscure an unwanted view. Numerous small woods in Yorkshire that appear to be plantations are, in fact, old or ancient woods which have been cleared and coniferised, or planted with a combination of beech and larch or pine. In 1794, Rennie, Brown and Shirreff, in their survey of agriculture in the West Riding, wrote: 'it is our opinion that larches and Scots firs would thrive in many situations, Wood of these kinds is much wanted'. A 'Hagg' or 'Frith' name, old boundary features or old deciduous pollards surviving at the edge of the wood will sometimes reveal its antiquity. The planting could take place on quite modest holdings or affect extensive areas. In the Ribble valley, for example, Lord Ribblesdale had more than a million oak trees planted at the end of the eighteenth century, as well as smaller numbers of ash, beech and elm. In the early nineteenth century, about 1,000 acres of land on the Barden estate in Wharfedale, which had previously been leased to tenants as wood pasture, were taken by the estate to become managed plantations. The changes

were initiated by Revd William Carr, incumbent of Bolton Abbey and agent to the Duke of Devonshire, in 1806–19. In 1810–11, 422,500 trees were planted at the Laund clearing, by the Wharfe, in former deer-park territory, about two-thirds of them being broadleaved and the remainder conifers. Selected deciduous woods were preserved to enhance the picturesque qualities of the riverside landscape, as at the Strid gorge. Carr created pathways and placed seats at vantage points offering views of the ruins of Bolton Priory and the Barden Tower, and the locality was visited and admired by arbiters of taste and exponents of the Picturesque, like Wordsworth, Ruskin and Turner.

The remodelling of countryside by the owners of great estates was made possible by the enclosure of communal land, whether by agreement amongst the tenants (the enclosure of land on the

Walled fields of various ages – and incongruous conifers – in this view of Wharfedale

Barden estates had mainly taken place in the second half of the seventeenth century), or by parliamentary enclosure. Wherever the old ways persisted, parliamentary enclosure brought their downfall. As noted above, local enclosures by agreement had already partitioned and removed the open fields and commons in many places; in others, however, the old countrysides retained many of their medieval layouts and communal arrangements. In places such as these, leading landowners would petition Parliament and, once the appropriate Act was passed and the commissioners had made their allocations, the public land, which might exist as strip fields, village greens, shared meadows or upland commons, could vanish entirely, with grid-works of walls or hedges tracing out the new patterns of private ownership. In many parishes, the contemporary fieldscape is a patchwork which is partly the creation of enclosure by agreement and partly

one of parliamentary enclosure. At Grassington, for example, several scattered strip holdings were sold to freeholders at the start of the seventeenth century; these farmers then swapped their strips around to create larger fields, whose elongated forms and curving margins still reflect their origins as plough strips. In 1792, parliamentary enclosure then augmented the privatised fieldscape, defining and prescribing the straight wall boundaries which partitioned the four stinted pastures between the existing gate-holders. Numerous Acts were passed in the latter part of the eighteenth century, and the process continued into the nineteenth century, with 600 acres being enclosed at Clapham in 1849 and Grinton Moor in Swaledale being enclosed in 1857. Of course, the nature of the land being enclosed varied greatly: in the Vale of York, the old open-field ploughlands bore the brunt of the transformations; in the Dales and Pennines, it was the upland commons that were enclosed. Normally, each Act would encompass a single parish, but the enclosure of the old Forest of Knaresborough in 1770–8 covered a massive area of 25,000 acres, spanning much of Nidderdale and lands beyond.

Parliamentary enclosure produced countrysides of straight walls or hedges and newly built solitary farmsteads. This is a view on Dodd Fell, near Hawes

Enclosure was a controversial and emotive subject. Since the awards given were notionally related to the size of the pre-enclosure holding, enclosure was welcomed by the more substantial farmers, who could look forward to receiving a conveniently compact new farm, and by large landlords, who could use the improvements associated with enclosure as a pretext for raising rents. The recipients of awards faced a hefty bill for hedging or walling their new, consolidated holdings, but this could be recouped. For example, John Ingilby at Ripley paid £195 for the Act, £658 for fencing, hedging and walling 384 acres, and £535 for new barns – but his rental rose by more than £100 per annum as a result. For countless small farmers, however, the costs of enclosure could not be found and their existence as private landowners might last for only a few weeks before they were obliged to sell up and leave. Enclosure spelt destitution or migration for the innumerable families whose existence depended on their ancient rights to pasture a few beasts and gather fuel and bedding on the common. Prior to the enclosure of the Forest of Knaresborough, a meeting in Hampsthwaite village alone produced a petition of opposition signed by 169 people, many of them probably the occupiers of squatter cottages who had every reason to fear the changes. Squatters who lacked any title to the lands that they occupied were often dispossessed, and small farmers who were left with unviable holdings and a bill for the obligatory fencing or walling were obliged to sell to their neighbours. One consequence of the migration of thousands of small semi-subsistence farmers was that more food was released for sale to the growing industrial communities: there must have been numerous wage-earning factory workers in Yorkshire who unwittingly consumed food that had originated on the very land that they had been obliged by poverty to abandon.

In the Dales, however, the enclosure of the upland commons did bring a reduction of overgrazing and increased the incentive to improve the pasture by draining and liming. Such commons might either be completely partitioned between private owners, or else they were surrounded by a wall and 'stinted', with each holder of a stint or 'beast gate' being allowed to graze a certain number of animals. (In some places, the commons and moors were already stinted, making it easy for the commissioners to make their allotments in proportion to the stints.)

In the Vales, the new hedgerow grid-works filled the void in the jigsaw of winding older hedgerows. Harris writes that:

> Parliamentary inclosure in the East Riding began with the inclosure of Catwick in Holderness in 1731, and thereafter slowly gathered momentum, until during the second half of the eighteenth century scarcely a year passed without the presentation at Westminster of at least one inclosure bill. Between 1730 and 1810 about 68,000 acres were inclosed in Holderness, 44,000 acres in the Vale of York, and 206,000 acres on the Wolds. By 1810 very little land remained uninclosed in the lowlands, though on the Wolds some 20,000 acres still awaited inclosure.

To the east of the Wolds, in Holderness and the Hull valley, the countryside was almost fully enclosed by the nineteenth century, though a few communities, like those of Mappleton and Kilnsea, would retain their open fields until the 1840s, and Brandesburton kept its common pastures; the reclamation of coastal land here was actively pursued when enclosure finally arrived, in 1846–7. In the North York Moors, the pattern of enclosure was patchier; much land had been enclosed by agreement, and the impact on the moors was slight. Spratt and Harrison attributed this variability to proprietorial factors and pointed out that: 'The relatively frequent parliamentary enclosures in the parishes between Pickering and Scarborough correlates closely with the old Forest of Pickering, where remnants of medieval forest laws and Crown ownership had for a variety of reasons inhibited the establishment of full-blown manorial townships; they tended to be owned by families wealthy and powerful enough to over-ride tenant opposition by resort to parliamentary act.' The effects were more obvious in the Dales, with straight walls slicing up the high commons and moors. (Some upland landscapes of dispersed laithe houses, geometrical fieldscapes and ruler-straight enclosure roads, like the one between

Brimham Rocks and Pateley Bridge, are very largely the creations of the late eighteenth century.) Below, there may be a pattern of newly enclosed rectangular pastures and older closes, defined by walls or hedges. Next, on the lower valley slopes, may be found the fields laid out across the old open village ploughlands, which had usually been completely enclosed by agreement before the days of parliamentary enclosure. The new provision of consolidated holdings could be a further inducement for farmers to desert the village for a new stone farmstead built nearer to the centre of each little empire (see below); even more noticeable features in the Dales, however, are the stone haybarns, dotted around the fieldscape and expressing a confidence in the new arrangement. In upland countrysides, enclosure

brought a frenzy of activity. On the commercial and social fronts, this involved the purchase of small holdings and the departure of many poorer folk, with those who could not find work as farm labourers leaving for mines and quarries or going further afield to the swelling industrial towns. On the land, it involved prodigious efforts in hedging and walling, the quarrying of additional stones to build the new access roads, and the erection of little kilns to burn the limestone needed to sweeten the pastures.

Throughout Yorkshire, parliamentary enclosure produced an exercise in boundary-making on an unprecedented scale. Those who acquired land according to the allocations of the commisioners were required to hedge it or wall it within the year. Traditionally, the choice between hedges and walls had been guided by environmental constraints concerning the ability of the land concerned to sustain hedgerow trees and the availability of cheap and durable resources of stone. Such considerations were noted by Harris in a quotation from an old account relating to the unsuitability of the chalk of the Wolds: 'The

Snow on old lead workings above Grassington, showing the superimposition of the earthworks of mining on the ancient 'Celtic' fields. Beyond, straight enclosure walls slice across the old commons

Stones upon the Place moulder like Lime in ye Winter, so that no Wall Fence can be made, therefore a Quick set must be raised whatever it costs', but protecting a young hedge was costly:'I am of opinion a good Ditch with very slight Rails and a good quantity of Gost or what is there called Whins [gorse] may nurse up a hedge . . .'. The unacceptability of the theory of hedgerow dating has recently been explained; one cannot deduce the age of a hedge by counting species. Other factors apart, medieval hedgerows were often composed of mixed seedlings gathered in the local woods; over time, certain very well-adjusted species, often elm or blackthorn, could invade a hedge, depressing the species count as time progressed. However, it is reasonable to suggest that hedges containing species like oak, crab apple, various wild plums, hazel and field maple are likely to be centuries old, while those dominated by hawthorn, with some roses, bramble, sycamore and elder, are likely to result from parliamentary enclosure, with the original hedging material having been purchased at one of the new nurseries. At the time of enclosure, walling was required on a scale that outstripped the capabilities of the local populations, and teams of professional wallers were employed. The Act would normally specify the location of quarries and of the access to them; the specification for the walls could also be quite specific and demanding. Raistrick has quoted the specification for the walls of the Grassington pastures in 1788: 'WE DO hereby Order and Award that the same shall be done by good stone Walls, in all places made 34 Inches broad in the Bottom and 6 Feet high, under a Stone not exceeding 4 inches in thickness, which shall be laid upon, and cover the Tops of the Walls in every Part, that there shall be laid in a Workman-like Manner 21 good Throughs in every Rood of Fence . . .'. The costs of walling were considerable, particularly to the smaller landholder. Mitchell writes that:

> When Studfold Moor, near Horton-in-Ribblesdale was enclosed in 1771, one of the landowners, Dr Wilson of Beecroft, paid £5 to the Enclosure Commisioners as his share of the cost. This charge was established at the rate of 10s for each stint that had been in his possession before enclosure took place. His share of the Moor consisted of two allotments totalling 42 acres. He arranged for the getting and leading of stones needed to make the flanking walls. This work lasted a year and cost the landowner about £25.

Arthur Young recorded that a waller working near Richmond in the mid-eighteenth century built 280 roods (two metre lengths) of wall which was 1.5 metres high in a year and was contracted at the rate of 5s. 6d. per rood.

The enclosure of common land was an essential precursor to the agricultural improvements which were spearheaded by the owners of the great estates. Williamson and Bellamy write that:

> The enclosure of the English landscape and the concentration of land in the hands of a small elite were accompanied by a fundamental revolution in attitudes to landed property. These new concepts of land and land ownership stimulated changes in the way that the countryside was used. In the Middle Ages there was no large-scale manipulation of the landscape in the display of personal status. The pattern of landholding would have made it difficult for any individual to make a large-scale alteration to the fabric of the countryside; furthermore, the nature of society was such that this was neither sought nor desired. It was only with the evolution of private property that the landscape began to be deliberately and extensively shaped for social and aesthetic purposes. For the first time the social elite displayed their power directly in the land.

One way in which this power was displayed was in the improvement and colonisation of low-value land on their estates. At Sledmere, an inscription on a rotunda proclaims:

> This edifice was erected by Sir Tatton Sykes Bart. to the memory of his father Sir Christopher Sykes Bart. who by assidulty and perseverence in building and planting and inclosing on the Yorkshire Wolds in the brief space of thirty years set such an example to other owners of land as has caused what was once a bleak and

barren tract of country to become now one of the most productive and best cultivated districts in the county of York AD 1840.

The tablet in West Heslerton church is more direct: 'Whoever now traverses the Wolds of Yorkshire and contrasts their present appearance with what they were cannot but extol the name of Sykes.' Like other powerful families in the east of Yorkshire, the Northcliffes, Osbaldestons and Legards, the Sykes had become deeply involved in agricultural improvements. Sir Tatton Sykes pioneered the use of bone as a fertiliser and, after the 1820s, mills where bones were ground were established in various towns and villages. He also developed a renowned flock of Leicester sheep on his estates, this breed becoming universal on the Wolds.

On the North York Moors, notable improvers included the stockbreeder Sir Thomas Dundas of Upleatham and Sir Charles Turner of Kildale, who devoted much energy to the reclamation of moorland. Kempswithen was the scene of Turner's most ambitious operation; Spratt and Harrison note that it was probably the relatively loamy soil which:

persuaded Turner that the ridge could be brought into productive use, not just for improved pasture or hay, but for cereal crops. To this end in 1773 he had the crest and both flanks 'pared' – vegetation cut down to ground level – and the parings burnt and spread. Radial drains were opened down the hillside to take off surface water, then large quantities of lime were carted in at great expense and ploughed into an area totalling many hundreds of acres. The whole was enclosed within

stone walls, lengths of which survive. Part was then laid down to grass and part to corn.

In the event, the plan was overambitious and the enclosure was abandoned after a decade or so, though Turner's scheme to improve Kildale enjoyed more success.

In 1836 the Yorkshire Agricultural Society was founded, with strong support from the region's aristocracy, like the Earl of Harewood, the Duke of Leeds and Lord Feversham. The progress of agricultural improvement could be charted by the displacement of local breeds and cross-breeds by others with superior qualities. The Dales had always had a reputation for hill sheep, but selective breeding with stock resembling the Scottish Blackface sheep resulted in the refinement of the Swaledale breed by the early twentieth century and of the Dalesbred by 1930. On the farms of the lower valleys, Wensleydale or Leicester rams were crossed with such hill sheep to produce Masham ewes, which could then be crossed with Suffolk rams to produce large, hardy lambs. The traditional cattle of the Dales were longhorns, but eighteenth-century breeders produced shorthorn cattle, dominating the region for a century or so; during the twentieth century, Friesian cattle almost evicted the shorthorn and Ayrshire breeds before encountering competition from a range of continental imports. The improvement of livestock did not monopolise the energy of estate owners in the Dales. The hunting-out of boar and then of deer engendered the pursuit of smaller quarry; during the nineteenth century, grouse shooting developed into a ritualised form of slaughter. Shooting on estates containing an extensive component of moorland may be operated by the owners or by syndicates. In the scenic and the

Gayle, near Hawes, was one of the centres of the old Wensleydale hosiery industry. It acquired a cotton mill in the 1780s, but then the textile industries slipped away to the coal-based factory sites

ecological senses, grouse shooting has considerable advantages; both the burning of the moor and sheep grazing favour the growth of soft new heather shoots which, in turn, favours the grouse. Even the excluded and disgruntled rambler must accept that the replacement of grouse moor by other land uses, like coniferous forest, would be a disaster for wildlife and country-lover alike. In the North York Moors, the proportion of moorland, within boundaries corresponding to those of the National Park, has fallen from 49 per cent in 1853 to 40 per cent in 1963 and 35 per cent in 1986.

If parliamentary enclosure allowed the social élite to manipulate the rural landscape, it also embodied potentials for change which could be exploited by the lesser tenants and freeholders. Where villages existed in Yorkshire, they were almost invariably the dormitories accommodating the labour force which could be found at work in the surrounding fields. The extinction, at a stroke, of communal farming massively undermined the *raison d'être* of the village; there was no longer any compelling reason why the farmer should not set up a new home with his family at the heart of their newly privatised holding. Harris notes that, on the Wolds:

> The new farmsteads frequently lay at some distance from the old sources of water in the village and their situation demanded the

If spared from afforestation, the countrysides of the dales are packed with interest. This typical view, over Skyreholme, near Pateley Bridge, towards Simon's Seat, shows old enclosure walls and fields having been invaded by moorland. The precise age of the farmstead sites is uncertain

provision of new sources of supply close at hand. To meet this need wells were sunk, sometimes to a great depth, and cattle ponds were constructed in the fields. Woldsmen were particularly skilled in these tasks and acquired a reputation throughout Yorkshire for the excellence of their field ponds, which were usually lined with puddled clay to prevent the escape of water.

More information about the situation in the Wolds before and after enclosure was provided by Gleave:

In 1770 the major part of [this] area was still being worked by means of open field system or some variation of it. Each township had its arable fields, worked in common, and its pasture which was frequently the highest area, most remote from the village nucleus. The homesteads and farmsteads were concentrated in villages and hamlets. Outlying farmsteads were the exception rather than the rule. . . . By 1850 parliamentary enclosure had taken place, affecting over 70 per cent of the area. The expanses of open arable and pasture had been replaced by smaller hedged fields and the Norfolk four-course rotation had been introduced at the expense of the earlier methods of rotation. . . . a rash of new outlying farmsteads, which originated in the period 1770–1850, reflects this new state of affairs.

Middleton-on-the-Wolds was enclosed in 1805; by 1818, there were 23 farmsteads still in the village and 7 located outside it, most of those in the village being associated with small allocations of land; in 1851, there were the 7 farmsteads outside the village, but only 14 remaining within it. The farmsteads located outside the village now controlled two-thirds of the acreage of the township. The dispersion of the farmsteads on the Wolds was associated with other changes; the sheep walks, many of them spread across lost village sites, were being ploughed up, and the Norfolk four-course rotation was replacing established systems of fallowing. Hayfield comments that:

'The enclosing and ploughing-up of the Wolds had three important consequences: First it led to a great increase in the manpower and number of working horses required to cultivate those lands. Second, it necessitated new and larger farm buildings to house the horses required to cultivate those lands . . . Third it required great increases in the amount of manure required to fertilise the fields.' These demands resulted in the development of a distinct class of buildings, the high barns of the Wolds, equipped with fold-yards for cattle, which, though much larger and far rarer, are the eastern equivalent of the field barns of the Dales.

In the period since enclosure, the principal changes affecting the farmed area of Yorkshire have been the wholesale destruction of wildlife and habitats as a result of 'scientific' farming and the Common Agricultural Policy, and a steady reduction in the diversity of production. This is apparent on individual farms, where the mixed farm, raising roots and cereals, cattle, sheep, pigs and a variety of poultry, is very much a thing of the past. It is also a phenomenon at the region-wide level, with many crops having been abandoned and relatively few new ones adopted. In 1817, working rabbit warrens are said to have occupied some 6,000 acres in the vicinity of Pickering alone. Rabbit farming had grown in importance in both the Tabular Hills and the Yorkshire Wolds in the early years of the eighteenth century, the skins being sold to hatters in Scarborough, Whitby, Pickering, Malton and York. A few of the warrens remained in use for much of the nineteenth century. Teazles were cultivated around a few lowland villages, the crop being introduced to the village of Biggin, near Selby, in 1770 and spreading in the surrounding district. The teazles were used to raise the nap on certain cloths and fabrics and fared better on the heavy clays of York than in the close vicinity of the textile manufacturing area. During the nineteenth century, most commercially grown chicory was produced within a few miles of York; by 1904, the English production was negligible and the market was almost entirely served by supplies from Belgium.

REFERENCES

J. C. Barringer, *The Yorkshire Dales* (Clapham: 1982).

T. W. Beastall, *A North Country Estate* (London: 1974).

H. M. Beaumont, 'Tracing the evolution of an estate township: Barden in Upper Wharfedale', *The Local Historian*, 26 (1996), pp. 66–79.

B. English, *Yorkshire Enclosure Awards* (Hull: 1985).

M. L. Faull and S. A. Moorhouse, *West Yorkshire: An Archaeological Survey to AD 1500* (Wakefield: 1981).

R. Fieldhouse and B. Jennings, *A History of Richmond and Swaledale* (Chichester: 1978).

M. B. Gleave, 'Dispersed and nucleated settlement in the Yorkshire Wolds', in D. R. Mills (ed.), *English Rural Communities* (London: 1973), pp. 98–133.

A. Harris, *The Rural Landscape of the East Riding of Yorkshire 1700–1850* (Oxford: 1961).

A. Harris, 'Chicory in Yorkshire: a crop and its cultivation', *Yorkshire Archaeological Journal*, 59 (1987), pp. 139–58.

A. Harris and D. A. Spratt, 'The rabbit warrens of the Tabular Hills, North Yorkshire', *Yorkshire Archaeological Journal*, 63 (1991), pp. 177–206.

C. Hayfield, 'Manure factories? The post-enclosure high barns of the Yorkshire Wolds', *Landscape History*, 13 (1991), pp. 33–45.

C. Hayfield and M. Brough, 'Dewponds and pondmakers of the Yorkshire Wolds', *Folk Life*, 25 (1986–7), pp. 74–91.

B. Jennings (ed.), *A History of Nidderdale* (Huddersfield: 1967).

R. A. McMillan, 'The Yorkshire teazle-growing trade', *Yorkshire Archaeological Journal*, 56 (1984).

W. R. Mitchell, *Drystone Walls of the Yorkshire Dales* (Giggleswick: 1992).

R. Muir, 'Hedgerow ecology and the landscape historian', *Naturalist*, 120 (1995), pp. 115–18.

R. Muir, 'Hedgerow dating: a critique', *Naturalist*, 121 (1996), pp. 59–64.

A. Raistrick, *The Pennine Dales* (London: 1968).

A. Raistrick, *The West Riding of Yorkshire* (London: 1970).

A. Raistrick, *Pennine Walls* (Clapham: 1988).

D. A. Spratt, 'Orthostatic field walls on the North York Moors', *Yorkshire Archaeological Journal*, 60 (1988), pp. 149–57.

D. A. Spratt and B. J. D. Harrison, *The North York Moors* (London: 1989)

T. Williamson and L. Bellamy, *Property and Landscape* (London: 1987).

CHAPTER SEVENTEEN

Market Towns and Routeways

IN THE COURSE OF THE INDUSTRIAL REVOLU-TION, the urban relationships of Yorkshire were transformed by the eruption of the gigantic Leeds–Bradford conurbation and the rise of several other important manufacturing centres, like Sheffield and Doncaster. With the exception of a very small number of purpose-built industrial settlements, notably Middlesbrough, however, the great industrial towns had all had previous existences as modest subregional market centres; beyond the industrialised areas, the rural localities have continued to be served by centres not transformed by industrialisation. Over time, the hierarchy of urban centres has experienced many changes as towns have expanded or lost their rankings. Today, York is somewhat dwarfed, if not overshadowed, by Leeds, although the dominance of the city in the urban affairs of the region in Roman and Dark Age times was almost absolute. York retained its position as the second city of England in the earlier medieval centuries and, though only comparable in size with a service centre like modern Skipton, it was still several times the size of any other town in Yorkshire. In the earlier part of the Tudor period, it was still one of the top six cities in the kingdom, with a population of perhaps around 8,000. From the 1460s to the 1560s, however, York was in the grip of recession and, despite the importance which its inhabitants had traditionally attached to maintaining its transport links, York's time as an international trading port and cloth-manufacturing centre was running out. To survive, York had to continue to attract immigrants; in the unhealthy medieval cities, death rates tended to exceed birth rates. Palliser notes that: 'The picture that emerges clearly is of a major city which, even in decay, could attract the young and hopeful not only from the surrounding settlements but from

distant towns and villages all the way to the Scottish border.' This frequent replenishment was essential, and long-established urban dynasties were unusual, as Platt explains:

> at York, too, there were other pressing reasons why the percentsge of newcomers among the city's freemen should have remained consistently high, even for the thirteenth century, in a period of population growth, it is unlikely that the citizenry could have maintained themselves without extensive immigration. One of the reasons for this weakness was a consistent failure of heirs, and it is a noticeable characteristic of English burgess dynasties wherever these have been examined, that they rarely enjoy a long life.

He pointed out that the medieval urban populations suffered from a reproductive weakness, and added: 'Undoubtedly, the most convincing explanation for this weakness is the high level of infant mortality in the towns . . . where [life expectancy] sank most cruelly in medieval England was in the earliest years of childhood, when resistance to bacteria had not had the opportunity to develop.' Until quite modern times, Yorkshire people were seriously underrepresented at medieval Parliaments. In 1536, the distribution of the main towns was roughly reflected in the demand, arising from the Pilgrimage of Grace, that a Parliament be held at York, with representatives to be chosen from Beverley, Pontefract, Richmond, Skipton and Wakefield; other market towns of some substance at this time would have included Knaresborough, Boroughbridge, Ripon, Hedon, Malton, Northallerton, Leeds, Halifax, Hull and Thirsk. All these places were still small and provincial in comparison to York.

Many of today's country towns must have emerged as small market centres in late Saxon times, although little is known about the infancy of such places. Towns with established Saxon pedigrees include Ripon, Otley, the royal manorial centres of Knaresborough and Wakefield, Pontefract (originally Tanshelf) and Conisbrough. Ripon was an important ecclesiastical centre from the time of Wilfrid in the seventh century, and local legend claims that Alfred the Great formalised its urban status in 886, appointing a wakeman, 12 elders and 24 assistants to govern the town. Ripon possesses a ceremonial horn which is held to have been a gift from King Alfred and, somehow, to constitute the charter; it also had 'a charter from Alfred's grandson, Athelstan, of 937'. However, Ripon's claim to be the oldest chartered city in Britain is scarcely flawless: the horn is undated and the charter apparently a medieval forgery, though perhaps based on a genuine original. There is no doubt that Ripon

was an influential early centre; the church here did receive charters from Henry I and Stephen, though the canons claimed already to have received the Athelstan charter, offering sanctuary from arrest within a one-mile radius of the church of St Wilfrid. The celebrations of the anniversary of 886 provided a reminder of the fragility of the claim to extreme civic antiquity and the lack of evidence that Alfred ever found his way to this remote part of his realm.

The clerks who compiled the Domesday evidence appear to have underestimated the urban aspect in Yorkshire. They recorded burgesses only at the Norman castle town of Pontefract, listed as Tanshelf, at Tickhill, where the 31 burgesses were listed under the old name of Dadley, at Bridlington and Pocklington. Doncaster was one omission. This was a significant market centre in the later Roman period, located at a road crossing and at the highest navigable point on the Don. The town will have engaged in the bulk shipping of goods like pottery and may have had an administrative role. The fort stood in the vicinity of the later castle, probably with an extra-mural civil settlement to its south, near the market place on the High Street. In the middle of the fourth

Terraced houses of the Industrial Age now nestle below the great Norman castle of Conisbrough

The Thursday market at Ripon. The market still flourishes, but its antiquity is currently the subject of some dispute

century, however, the town ceased to function. A revival probably took place in the late ninth century and, by the early tenth century, Doncaster was operating as a town again. A form of urban resurgence might have taken place much earlier, for the entry for 764 in Simeon of Durham's history lists Doncaster in the distinguished company of Winchester, Southampton, London and York as places destroyed by fire. Scarborough also appears to have had some kind of urban existence at the time of the conquest, for its people were sufficiently numerous to make a stand against Hardrada's invading forces. The precise location of the early settlement here is uncertain, and Farmer writes: 'since 1979 archaeological finds (unfortunately unstratified) on the sheltered flat plateau of land just below the castle entrance and close to St Mary's church indicate settlement in this area at an early date in the 11th century, raising the possibility that it was this "town" that was burnt by Harald Hardrada in 1066 and not the area by the harbour.' The Norman defences of the headland began with the building of a wall in about 1130, and Henry II built the royal castle in 1158–69. The town then developed below the castle, in the vicinity of the harbour and St Mary's church, and achieved borough status in the 1160s. The old town then expanded to include Newborough. Scarborough, which supplied the numerous abbeys of Yorkshire with sea fish, was one of the first seaports to develop a commercially significant fishing industry.

During the twelfth century, there was considerable population growth and a quickening of the economic pulse of the countrysides as communities recovered from the Harrying. Towns developed beside several Norman castles in Yorkshire, and Hey writes that, in addition to Tickhill:

Successful and not so successful towns were also founded near to Norman castles at Almondbury, Harewood, Helmsley, Kirkby Moorside, Knaresborough, Malton, Northallerton, Pickering, Richmond, Scarborough, Sheffield, Skelton, Skipsea and Skipton, and of course in all other English counties as well. Sometimes, however, the topography of a medieval town is rather more complicated. At Wakefield the town grew around the ancient centre marked by the parish church, well away from Sandall castle, and at Thirsk the towering Perpendicular church of St Mary and the nearby site of the former Mowbray Castle are remote from the busy market town on the opposite bank of the Cod Beck. Presumably, church and castle were the focal points of the original settlement, the Thirsk whose name is derived from an Old Norse word for fen. However, by 1145 at the latest a Norman town had been established around the enormous rectangular market place at the other end of Kirkgate.

The earliest history of Knaresborough is uncertain, its name suggesting a possibly pre-conquest stronghold on the rock overlooking the gorge of the Nidd. The Normans might have used Knaresborough as a base during the Harrying; by the end of the eleventh century, the royal Forest, covering about 100,000 acres, formed part of the new Honour of Knaresborough. The first surviving reference to the castle dates from 1130, and the castle served as the administrative centre for the forest. A market centre defended by a ditch developed beside the castle, with bond tenants living in Bondgate, outside the ditch.

By the middle of the thirteenth century, Yorkshire probably had 40–50 towns. In the human world, patronage was essential for advancement in the Middle Ages; to gain a foothold on the ladder to urban success, a village, too, needed an influential patron, either an enterprising manorial lord or a notable churchman. Settlements developing in the shadow of a castle or beside a religious community were boosted by such associations. The key to possible success involved the acquisition of a licensed market, and sponsors could expect to benefit from the resultant commercial activity by the collection of tolls and fines from market trading. They were also able to enjoy access to a convenient source of trade goods and to exploit the outlet for the products of their estates. Lords who could secure the right to hold annual or periodic fairs in addition to their weekly markets could expect to attract itinerant merchants, bringing the more exotic or specialised goods which circulated on the international trading networks and which were otherwise unobtainable in the locality. In some cases, the appropriate royal charter only regularised a market which could have existed since well before the Norman conquest; in others, the market and commercial settlement were the artificial creations of noble entrepreneurship. Existing centres which became the capitals of great feudal honours had a head start in the world of commerce. Richmond, basking under the patronage of its earls, gained a market charter during the first half of the eleventh century, and Skipton, Conisbrough and Pontefract were other particularly favoured Norman castle towns. A village operating a two-field system appears to have existed at Skipton before the conquest; after the manor was granted to de Romille, a rock defined by the steep gorge of a meltwater channel was selected as his castle site and the headquarters of his new honour of Skipton, a market charter was obtained, and the settlement developed on the sloping ground below. Raistrick writes that, in 1311: 'The Castle was then the centre of administration of a large forest area and the site of the Honours Courts and was a very busy place. In the fourteenth and fifteenth centuries the growth centred around the market place and along Sheep Street. The town became stabilised along the High Street leading down from the Castle and church to the line of the Roman road at the foot . . .'. This pattern survives in its essentials, though the building frontage line has trespassed on the market area. Raistrick noted that: 'In 1379

the town had seventy-nine labourer's families and forty-eight tradesmen and craftsmen, a large proportion of them merchants and traders. These included thirteen families concerned with making or dealing in cloth, and Peter de Brabant, his sons and their wives, Flemish weavers, were making shalloons. There was a corn mill and a fulling mill, inns and shops.'

A settlement of Tanshelf was recorded in 947, when Archbishop Wulfstan of York swore fealty to King Eadred there at a crossing on the Aire; Domesday recorded Tanshelf as a royal manor with 60 burgesses. This original manorial centre was superseded in the twelfth century, when a new town was laid out to the west of the castle, with Micklegate as its broad market street and its limits marked by Walkergate and Back Northgate; this new town was Pontefract, 'broken bridge' in Norman French. The Micklegate market, which was chartered in 1194, was so successful that a second borough of West Cheap was created nearby and chartered in the 1250s. The development of the West Cheap market, just outside the town wall, provided a much more spacious trading zone in the area between Salter Row, Beast Fair and Market Place. Here the street names betray the trading functions of the market before streets encroached across its area: Corn Market, Wool Market, Shoe Market, Horse Fair, Salter Row, and so on. The new town flourished; in 1377 it had 1,085 taxpayers and would have ranked third in Yorkshire, after York and Hull. The search for the location of Tanshelf made a significant advance in 1985, when a Saxon church and burial site were excavated on sites just to the south-east of the castle.

Of the larger towns of industrial Yorkshire, Sheffield existed as an unexceptional market town at the close of the medieval period; at its start, Sheffield was a village on a ridge near the junction of the Don and the Sheaf. A small castle was built on a hillock there in the twelfth century, and the street names Castle Green and Mill Lane relate to the setting of the village. Seventeenth-century Sheffield had two weekly markets and two annual fairs. There were probably around 3,500 townspeople; 2,207 residents were listed in 1615. The castle was destroyed under an edict of

Cromwell in 1648; the market place adjoined the castle site and had stalls, the town bakehouse and at least 17 shops; a further 11 shops were situated beneath the Town Hall on the High Street. Kingston upon Hull began as a humble settlement whose name, Wyke, denoted its stuation upon a creek. Edward I obtained the manor in 1293, which had already been developed by Cistercian monks. The monks of Meaux Abbey had diverted the channel of the River Hull, dug a straight course to provide a more navigable link with the Humber estuary, and established the outlines of a new, rectilinear town plan, a grid-work of streets laid out to the west of the earlier monstic nucleus. Edward appreciated that Hull had a strategic location as a halfway house on the invasion route to Scotland, and the monks were persuaded to exchange their port for other lands in Holderness. By the close of the thirteenth century, the king's town was the third seaport in England, and 55 tenements in the property grid were occupied. Under Edward's patronage, the incoming roads from York and Beverley were improved, the harbour installations were enhanced, the fairs and markets extended, and, in 1331, the townspeople gained complete independence from feudal controls. In the course of the century, this independence was proclaimed by the construction of a town wall of brick, with 21 towers. The wall was destroyed after 1767 to create new docks, but excavations have revealed the layout. Hull controlled the river traffic into the interior of Yorkshire as well as the traffic of the Trent. Small seagoing craft could still navigate the Ouse to the riverside quays at York, but the city became increasingly dependent upon Hull for the conduct of its international trade. Links with Sheffield existed via the River Idle and the town of Bawtry, a small commercial centre in medieval times, usefully situated beside the bustling Great North Road, and the successful creation of Robert de Vipont in 1213.

The nature of the connection between *Loidis*, the presumed focus of the British kingdom of Elmet, and Leeds is uncertain. In 1207, Maurice de Gant established a new town on his manor at Leeds. It was attached to an existing village occupying the Kirgate locality, and the essential framework of the Norman new town is very prominent in the layout of modern Leeds, with Briggate, where the main market activity was conducted, leading up to Headrow, where a row of dwellings must have stood at the head of the settlement. Leeds had a 'T'-shaped layout, therefore, with Briggate as the stem and with attenuated burgage plots running back from each street. For centuries, Leeds had a rival in the great manorial centre of Wakefield, with Leeds emerging as the superior candidate as the regional capital in southern Yorkshire in the seventeenth century. Pre-industrial Bradford was merely a village.

The comprehensive form of medieval planning that was evident at Kingston upon Hull, Pontefract and Leeds was not unusual. At Hedon, the carefully planned origins of the town are still plainly evident; its square form was delimited by a boundary ditch and, inside, the area was set out with streets running from north to south. The borough was planted by the twelfth-century lords of Preston in Holderness, who cut three canals from the Humber tributary stream, the Hedon Haven, to enable shipping to gain access to the middle of the settlement. In 1203–5, Hedon ranked eleventh among English ports, but its commerce soon declined as the stream channel silted. Only around a third of the designated area is occupied today, and only one of three medieval churches survives. Boroughbridge is the planned product of a shift of growth and trading from the Roman centre and Saxon royal manor at Aldborough, nearby, following the construction of a new bridge across the Ure here in 1145. The streets were set out to a grid-work pattern around the market place. Borough status was achieved twenty years after the foundation of Boroughbridge; the one parish then contained two boroughs and subsequently sent two members to Parliament, the seats becoming pocket boroughs of the Duke of Newcastle. A considerable measure of planned redevelopment must have been involved in producing the layouts of most market towns, for market areas like the square at Ripon, the great rectangle at Masham – now almost empty, even on market day – the triangle at Bedale, and several other examples could not have appeared by accident. Either such market places

The planned medieval layout of Hedon, fossilised by the decline of the town as the result of competition and the silting-up of its water link to the sea. (Cambridge University Collection)

were original features of planned foundations, or else ample space for trading was obtained by the demolition of pre-existing buildings. Some places had more than one market, like Middleham, chartered in 1387. Brown comments that:

Middleham is a medieval town, certainly in terms of its layout which has changed little over the centuries. The market place and cross, the symbol of 'honest dealing', date to the fifteenth century as do the Swine market and its monument. The dual market places are typical of medieval towns, the lower market being used for general trade established by the market charter; whilst the Swine market was the site of trade in pigs, one of the greatest of its kind, making full use of the wooded land surrounding Middleham.

The infilling of a market area, initially perhaps by the building of permanent shops to replace the stalls of the open market, reduces and distorts the original layout; at Richmond, the market would have filled the semicircular outer bailey overlooked by the castle; at Knaresborough, the medieval trading area has contracted to a modest L-shaped space. In a few cases, however, a failure to expand and flourish leaves much of a medieval layout intact: at Hedon, competition from Hull and the lost port of Ravenser-Odd and the silting of its link to the sea brought the decline of the town and, consequently, the fossilisation of its layout. At the start of the nineteenth century, Hedon was a rotten borough with just 300 electors. Sheriff Hutton was originally just Hutton, until Bertram de Bulmer, the Sheriff of York, built his motte-and-bailey castle there in 1140. It was not until 1378 that a market charter was granted, when a large, square market green may well have been created by demolishing a number of roadside dwellings; it caused a shift of the focus of the settlement away from its original nucleus of small green, church and castle. The new quadrangular palace castle was built beside the green in 1382.

As most older markets prospered and the more obvious commercial prospects were exploited, so entrepreneurial lords were tempted to establish trading in various less promising settlements. Some later markets and fairs, like Middleham, Pateley Bridge (1320) and Ripley (1357), were reasonably successful, the village of Ripley probably having been moved by its new owners to exploit a better commercial site at about the time that the charter was obtained. Some, like Hampsthwaite (1304), linger as small annual feasts or fairs, but several, like Carperby, in Wensleydale (1305), had little potential, even though a cross may endure as a reminder of the long-dead aspirations. The competitions between urban markets could be quite intense, and memories of privileges bought or granted were long. Men of the honour of Knaresborough claimed the right to exemption from market tolls and made a nuisance of themselves at fairs at Otley and Ripon: when the Archbishop of York hired soldiers to keep order at Ripon fair in 1441 they were ambushed on their way back to York and two of them were killed by the Knaresborough hooligans. Rivalry between the different village markets was intense; in places like Wensleydale there can never have been sufficient trade to support all the contenders.

The commercial activities of old Yorkshire depended on a system of roads and bridges: the roads were generally inadequate, and the bridges – or lack of them – reflected the perennial problem of finding benefactors to support their construction and maintenance. Prehistoric roads and trackways can seldom be identified, but it is perfectly likely that scores of winding country lanes

The superb fifteenth-century bridge across the Tees at Croft must have made a vital contribution to late medieval transport. Much later, Croft attempted to develop a role as a spa, but was soon eclipsed by the rise of Harrogate

in the Dales and elsewhere date back to Iron Age times. Such tracks would have supplemented the skeletal network of engineered new roads that the Romans provided and, in due course, the medieval market towns were sustained by roads which had developed spontaneously under the stimulus of market trading or which might have been forged in Dark Age times. Dark Age or early medieval dates are often attributed to the old salt ways of Yorkshire. Some of these roads still retain their old 'Saltergate' names, though the association with the vital salt trade has sometimes been forgotten and the meaning of the name has been confused, as with the so-called 'Psaltergate' from Otley to Blubberhouses, in the Washburn valley. Like some of the other roads that were heavily used by the monks, salt ways are mentioned in early medieval charters, when they might already have been ancient, although Saltergate, on the North York Moors, running inland from old coastal salt-pans, escaped mention until 1335.

Most roads are difficult to date; one may not know whether the maze of back roads with names like Arkendale Road, Minskip Road, or Wath Lane developed to link existing settlements, or whether the settlements developed along existing lanes (the latter being, perhaps, the more likely option). The introspection of the medieval villager has been exaggerated; even if most trips extended no further than the nearest market, long-distance routeways were needed to link abbeys to their furthest estates; to serve the retinues of humble merchants moving from market to market, or the greater traders plying the network of fairs; to allow lords to move between their far-flung estates; and to serve the roaming dealers in wool. Occasionally, even a villager might travel far afield, perhaps being despatched to a seaport to buy a barrel of herring for the manor, or visiting a distant shrine. Packhorse roads in the Dales could be remarkably direct, but arduous; as traffic increased and wheeled transport gained in importance, a lower, gentler, but more devious line would be favoured. Some, at least, of the 30 or more crosses which punctuate the North York Moors were probably erected as waymarkers in the early medieval period to guide travellers on the bleak or snowbound trackways; at various

places in the Pennines, rough-hewn gritstone pillars marked the track.

Another type of long-distance routeway developed soon after the close of the medieval period, as relations with Scotland became less volatile (the citizens of York had been at liberty to kill any Scots found within their city walls). The trade in Scottish cattle became very important during the seventeenth century, the cattle being driven across the Southern Uplands and along the Pennines from places as far afield as the Highlands and Islands. The drove-roads avoided the cultivated areas and traversed the high plateaux, where track-side grazing could be found. The beasts were bought at 'trysts', held at places like Dumfries and Falkirk, and then resold at great gatherings to English graziers for fattening. Malham had one of the great gathering grounds of the droving trade, and thousands of cattle were handled by the markets at Masham, Ripon and Skipton.

Those people less hardy and more encumbered than the drovers were perpetually scandalised by the state of the country's roads, though generally unable or unwilling to do much to ease the situation. In the 1720s, Daniel Defoe described how 'From the Wharfe we went directly north, over a continued waste of black, ill-looking, desolate moors, over which travellers are guided, like race horses, by posts set up for fear of bogs and holes, to a town called Ripley.' Even so, he was impressed by river bridges in the Dales: 'no part can show such noble, large, lofty, and long stone

---

Whitby has a more distinguished history than most other tourist resorts. The Romans had a signal station here, and St Hilda founded her abbey in 657; seven years later, the synod of Whitby changed the direction of Christian worship in the North. After the Reformation, the medieval abbey church existed only as a sea-mark for sailors on the North Sea (becoming a target for the German fleet in 1914). Whitby's importance rose again, in the eighteenth century, as a whaling post; in Victorian times, the jet-carving industry enjoyed a boom here. Now, in seeking its future as a tourist centre, Whitby can profit from a very genuine charisma and a remarkable history

bridges as this part of England, nor so many of them'. Almost a century later, the conditions for travel between two such significant centres as Leeds and York could be treacherous, and in 1708 Ralph Thoresby confessed to his diary: 'We found the way very deep, and in some places dangerous for a coach that we walked on foot, but the Lord preserved us from all evil accident, that we got to our journey's end in safety, blessed be God.' An answer to the inadequacy of many important stretches of road was soon discovered in the creation of turnpike trusts, which relieved parishes of their neglected responsibilities, improved the highways and obliged travellers to pay to help in recouping the costs. The first Yorkshire turnpike received parliamentary approval in 1735 and ran from Elland and Halifax over Blackstone Edge – near a surviving stretch of Roman road – to Rochdale. Smith quoted the strict specification laid down for the Richmond–Lancaster road of

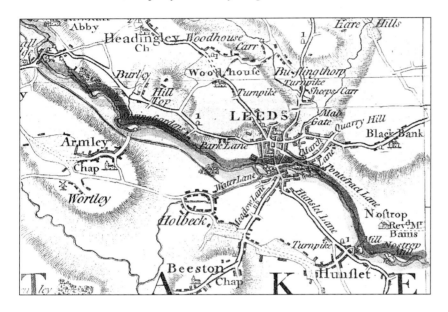

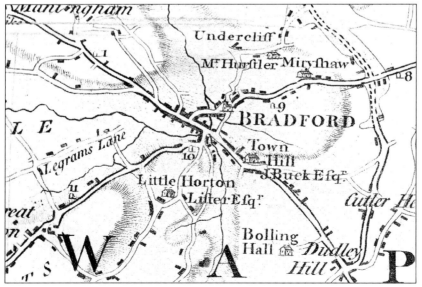

Leeds and Bradford as they existed in the 1770s, with the Industrial Revolution just a few decades old. Leeds and its suburbs now cover all but the south-eastern corner of the map, extending far beyond its borders to the north, west, south and east. Bradford extends far beyond the margins of the map in all directions

1754: 'The road is to be casten six yards broad within the trenches, to be well formed and as near as can be Levell taking down ridges and filling up hollows. To be stoned four yards broad and ten inches thick. To be very small brocken and well covered with the best gravill and earth, conduits to be made where necessary . . .'. There were still many who preferred wading in mud to the paying of tolls; in 1753, the Leeds mob walked out to destroy the toll-gates at Harewood and had a massive affray with the estate tenants. In the aftermath of the fight, troops from York killed eight members of the mob who were continuing their riot in Briggate.

Most people, however reluctantly, could recognise the physical and commercial advantages of the turnpikes. In 1066 it had taken King Harold a week to march his army from London to York; in 1700 it still took the traveller about eight days to get from York to London by waggon; by the end of the century, however, one could complete the journey in two days. In 1785 the first Royal Mail coach made the Leeds to London run in 26 hours. The West Riding of Yorkshire was a focus for intense and violent opposition to the turnpike movement with a tollgate at Selby being destroyed in broad daylight in 1752. Freeman wrote that: 'The destruction of the Selby gate in 1752 is clearly not inconsistent with the opposition which had been forthcoming from the rural communities around the lowest reaches of the Aire and Ouse rivers roughly a decade before.' That farmers were generally among the antagonists is certainly plain from a ditty that marching rioters are reported to have chanted:

> Corn, coals and lime shall go free
> Or else no turnpike there shall be.

In June 1753, 12 gates and 6 gatehouses were destroyed around Leeds, and similarly violent action was widespread in the West Riding. But there were many more who came to recognise the advantage of having efficient roads. Soon after the enclosure of the Forest of Knaresborough, in 1770–8, a turnpike was built to link Knaresborough and Skipton; its surveyor then remarked that, as a result: 'though scarce a single cart was before seen in the market of Skipton, not less than 200 are weekly attendant on that market at present'. Much shorter stretches of improved road were made in the course of parliamentary enclosure, but since the awards covered only a parish, the straight new enclosure roads appeared in a piecemeal and unintegrated manner.

REFERENCES

P. Armstrong, 'The old town of Kingston upon Hull', in D. Symes (ed.), *North Humberside: Introductory Themes* (Hull: 1978).

M. Aston and J. Bond, *The Landscape of Towns* (London: 1976).

M. W. Beresford, *New Towns of the Middle Ages* (London: 1967).

S. V. Brown, *Castle, Kings and Horses* (Hawes: undated).

R. Davis, *The Trade and Shipping of Hull, 1500–1700* (York: 1964).

D. Defoe, *A Tour through the Whole Island of Great Britain* (London: 1962).

J. F. Edwards and B. P. Hindle, 'The transportation system of medieval England and Wales', *Journal of Historical Geography*, 19 (1993), pp. 123–34.

P. G. Farmer, 'Early medieval settlement in Scarborough', in T. G. Manby (ed.), *Archaeology in Eastern Yorkshire* (Sheffield: 1988), pp. 124–48.

M. Faull and S. A. Moorhouse (eds), *West Yorkshire: An Archaeological Survey to 1500* (Wakefield: 1981).

M. Freeman, 'Popular attitudes to turnpikes in early-eighteenth-century England', *Journal of Historical Geography*, 19 (1953), pp. 33–47.

E. Gillett and K. A. MacMahon, *A History of Hull* (Oxford: 1980).

D. Hey, *Packmen, Carriers and Packhorse Roads* (Leicester: 1980).

D. Hey, *Yorkshire from AD 1000* (Harlow: 1986).

A. Kellett, 'King John in Knaresborough: the first known royal Maundy', *Yorkshire Archaeological Journal*, 62 (1990), pp. 69–90.

J. Langdon, 'Inland water transport in medieval England', *Journal of Historical Geography*, 19 (1993), pp. 1–11.

J. R. Magilton, 'Tickhill: the topography of a medieval town', *Transactions of the Hunter Archaeological Society*, 10 (1979), pp. 344–9.

D. M. Palliser, 'A regional capital as magnet: immigrants to York 1477–1566', *Yorkshire Archaeological Journal*, 57 (1985), pp. 111–23.

M. S. Parker, 'Some notes on the pre-Norman history of Doncaster', *Yorkshire Archaeological Journal*, 59 (1987), pp. 29–43.

C. Platt, *The English Medieval Town* (London: 1976).

S. Porter, *Exploring Urban History* (London: 1990).

A. Raistrick, *West Riding of Yorkshire* (London: 1970).

G. Scurfield, 'Seventeenth-century Sheffield and its environs', *Yorkshire Archaeological Journal*, 58 (1986), pp. 147–71.

C. C. Taylor, *Roads and Tracks of Britain* (London: 1979).

A. Wilmott, 'Pontefract', *Current Archaeology*, 106 (1987), pp. 340–4.

G. N. Wright, *Roads and Trackways of the Yorkshire Dales* (Ashbourne: 1985).

Yorkshire Archaeological Trust, *2,000 Years of York: The Archaeological Story* (York: 1978).

CHAPTER EIGHTEEN

Into the Industrial Revolution

THE PARLIAMENTARY ENCLOSURE of common lands and the turnpike movement paved the way for the Industrial Revolution, although the 'Revolution' was not a single traumatic event, but an accumulation of changes which developed over a period of several decades. As Gregory remarks, it is easy:

> to forget that the changes which were involved – and not all of them were benign – occurred unevenly in both space and time: that the structure of the space – economy was not transformed over night and that its rhythms of production did not accelerate in anything like a continuous way. Qualifications like this are important because they draw out the subtle textures of the past as well as its more dramatic patterns and, in particular, because they remind us of the diversity of regional experience during this period.

The changes that ensued during the decades of intense industrialisation and urbanisation affected much more than the economy, the landscape and the structure of society: they had profound cultural and psychological consequences. They had the effect that the village ploughman or shepherd living in rural Yorkshire in the mid-eighteenth century might have seemed far closer, emotionally and culturally, to the Neolithic herdsman and cultivator of 5,000 or 6,000 years ago than he did to his grandchildren, who would live in courts or basements in a northern mill town. Even so, these grandchildren took with them elements of the old peasant ideology, which would gradually be reshaped into urban/industrial values. Thus, workers in the Nidderdale flax and linen mills in the 1840s who had the use of a patch of pasture could subscribe to a cow club, and Jennings records how: 'Members paid annually sixpence in the pound on the value put on a cow, and if the cow died through sickness or accident three-quarters of this value was paid in compensation ... Between 1846 and 1892 the club paid out £528. 10. 0 in compensation and spent £185. 11. 0 on vetinary help.'

Today, more than two centuries after the commencement of the Industrial Revolution, the images of the period are still charged with emotion, and one must remember that our perceptions need not correspond with those of the people who experienced the events. To quote from Gregory again:

> To the labourer trudging across the fields on a winter morning, or the clothier leading his pack horse along the lanes to Leeds, discussion about concepts of utility or value had precious little relevance; in making sense of themselves, their communities and their relationships with the outside world they drew on a set of usages which, for all their differences, were enshrined in common rather than academic argument, and which were embedded in the day-to-day practice of what Thompson has called a 'moral economy' rather than in the more abstract promise of a political economy. The culture to which they subscribed reached back to – and was in large part an affirmation of – a parochialism and paternalism: a vision of the world in which men were bound to the earth on which (and through which) they lived, and a model of society in which men were bound to one another by a mutually reinforcing set of social obligations.

In due course, the people emerging from this remarkable episode of industrialisation (which we can now recognise as having ended recently and having yielded to an uncertain post-industrial experience) did so with their numbers hugely increased and with their regional identities as Yorkshirewomen and -men strengthened. The eruption of industrial towns like Barnsley, Dewsbury, Huddersfield and Batley had added an urban dimension to the Yorkshire character, and popular perceptions of 'Yorkshireness' had come to relate more strongly to the urban/industrial qualities than to the rural aspects. The industrial experience produced a ravaged landscape in places, though in fact, contrary to the mental stereotypes still current further south, only a fraction of the area of old Yorkshire was transformed. Historic countrysides survived in many places; even the ugliness of the working townscapes of the nineteenth century, as viewed today, is tinged with a strangely compelling dignity and power. Meanwhile, new generations of Yorkshire people, lacking any ancestral history as peasants or tenants in the Dales, Vales or Wolds, are emerging amongst the Pakistani, Indian, West Indian, Chinese, Italian and East European communities; in so doing, they are adding another chapter to the history of ethnic diversity, which includes Celts, English, Norsemen, Danes, Flemings and members of several other older cultures whose names are not even known.

The Industrial Revolution and its expression in the landscapes of Yorkshire could not possibly be covered in one chapter, nor even explored in any depth in a whole book of the length of this one. Rather than attempting the impossible, therefore, the emphasis here has been to outline the path into industrialisation from what had been a series of essentially rural settings. In Yorkshire, industrialisation did not tend to precipitate people from the communal meadow to the assembly line in a single step; the establishment of shift work and factories tended to be preceded by a well-established phase of domestic industry, involving households which combined the roles of smallholding and cottage-based manufacturing. The great majority of farmers in the Dales were the tenants of very modest holdings, and the ability of a family to survive often depended on secondary occupations. Even then, the levels of living could be desperately low in relation to modern expectations. Wills and probate inventories reveal the asceticism of cottage life: when John Luty, a weaver of Clint, in Nidderdale, died in 1686, he left a mare, a colt, two cattle and four calves, eight sheep and a little pig, and fifteen shillings worth of sown corn. His neighbour, James Speakman, died in the same year; he had called himself a yeoman, though all he owned was a couple of horses, a cow and a calf, a pair of geese, four hens and a peck of grain and bits of mill gear valued at 37 shillings. Robert Thurcross had died nearby in 1624; his most valuable possession was eleven loads of marl and manure. It is clear, when more detailed inventories are available, that many country people were part-time textile workers, for spinning wheels, yarn, cards and lengths of cloth were left, as well as the stock and paraphernalia of farming. Thus, when William Hardcastle died in 1681, his house was littered with the relics of his work as a dairy farmer and linen weaver:

Purse and apparel £2; *House*, three chairs, two buffet stools, one iron pot, one little kettle, two pans, five pieces of pewter, some cheese and other hustlement 13/4;
Parlour, one bed with bedding on it, one cupboard with some other hustlement 10/-;
Milk House, one churn, two barrels, six cheese vats, certain milk bowls and shelves and other hustlement 6/8d;
Chamber, two chests, two beds with bedding on them, two forms, one doughtrough, one kimlin, 1 ark, and other hustlement 1 pound;
Working house, one loom with gears belonging to it, one cheese trough, one cheese presser and other hustlement 10/-;
Oven house and turf house, certain [?] and one harrow and other hustlement 10/-;
Barn, certain hay and oats and beans 5 pounds; 3 cows, 2 young heifers, 1 calf 6 pounds; 2 mares, 7 sheep, 4 geese, 3 hens £3; Total £19-10- 0.
Debts owing by deceased: Funeral expenses 3 pounds; other debts £15.

Industry had complemented farming in the semi-marginal agricultural environments of the

Dales since the Middle Ages, if not before, and a number of activities had been pioneered by the old monastic landowners. Earlier, the Romans had mined lead at Greenhow, overlooking Nidderdale, and perhaps in Swaledale, too; early in the Middle Ages, Fountains and Byland Abbeys exploited lead, iron and charcoal resources in Nidderdale, and Jervaulx Abbey worked iron and lead in Swaledale and Wensleydale. On the North York Moors, Guisborough Priory was involved in the making of iron in Glaisdale, Byland Abbey worked iron in Rosedale, and Rievaulx Abbey had a forge in Bilsdale. Coal was mined by Bolton Priory in Wharfedale and by Jervaulx Abbey in Coverdale. During the 1560s, Elizabeth I had encouraged German miners to exploit the minerals of Cumbria, and this introduction of new techniques accelerated the development of the mineral resources of the Pennines. Some rich landowners acquired mining empires, with the Dukes of Devonshire controlling lead mining in Wharfedale. The combination of occupations within the cottage economy was again evident, but the relative importance of mining (or quarrying) and farming varied from household to household. Lead miners would often supplement their livelihood with the produce from smallholdings, as exemplified by an agreement made in Chancery to settle a dispute in 1613, which allowed the construction of dwellings for lead miners near the workings at Greenhow, high above Pateley Bridge, at an altitude of more than 385 metres: 'notwithstanding the common pasture to the tenants also to be granted . . . there may be cottages erected for the miners and mineral workmen on the said waste, and some competent quantity of ground to be improved of the said waste and laid to them, and also for the keeping of draught oxen and horses for the maintenance of the mines, always leaving the tenants sufficient common'. The habit of dual economies persisted in the Dales until the later phases of the Industrial Revolution. Jennings notes that:

> Five of the eight farmers who lived at Middlesmoor in 1851 had other occupations. Two were innkeepers, the others a blacksmith, a carrier and a postman. The greatest concentration of small farms was found in areas where the linen and lead mining industries were, or had been, strong. The decline of linen weaving in Bishopdale and other places during the middle decades of the nineteenth century did not cause an immediate change in the pattern, as a smallholder's family might find employment in a flax mill, and there were other crafts which could be combined with farming, e.g. carpenter, wheelwright, cobbler. The influence was greatest in Greenhow, a village which had come into existence not as a farming but as a lead mining community. Virtually all the land was held as farms or small-holdings by the miners.

Other industries, like quarrying, coal mining and iron working, were significant locally and had long histories, but the textile industry had been important throughout Yorkshire for centuries before the Industrial Revolution. Yorkshire had mainly specialised in the production of raw wool for export, but, during the fifteenth century, the region began to rival East Anglia in the manufacture of cloth. By this time, Yorkshire was able to exploit the experiences it had gained in the manufacture of textiles; even in the 1370s, the little township of Appletreewick, in a backwater of Wharfedale, supported several weavers and a dyer, and Ripon had 38 textile workers. While spinning and weaving were largely cottage industries and would remain so for centuries to come, Yorkshire enjoyed enormous resources of water power, and a multitude of streams were harnessed to drive the fulling mills, where fuller's earth was beaten into the newly woven cloth; the earliest fulling mill was recorded at Ripon in 1184. In the post-medieval period, the industry gradually drifted westwards from centres like York and Beverley to the Leeds area, where water resources were more plentiful and the restrictive practices of the industry were less entrenched. By 1724, Leeds was said to have the largest cloth market in England. During the Middle Ages, monastic buyers had ranged far and wide to purchase the wool from the farmsteads and, subsequently, they had counterparts in the itinerant 'broggers' of Halifax. Some of the wool purchased was carded and spun as a farmstead industry, and some of this was sold to knitters rather than weavers. During the sixteenth century, knitting became an important industry in and around Richmond and Hawes, while Dentdale developed a specialisation in knitting knee stockings for the army during the eighteenth century. There were numerous people who had triple occupations as farmers, miners and knitters. Jennings's study of wills and inventories from the forest of Knaresborough for the period 1551–1610 shows that 86 out of a total of 207 inventories mention textile equipment, yarn or fibres. At this stage, wool was more important than flax and hemp; of the 25 looms recorded, 21 were used mainly for weaving some form of woollen cloth. Those without looms were still likely to be involved in the textile industry, for: 'The households without looms probably put out their yarn to be woven by "customer weavers", who wove a customer's yarn into cloth, very much as a dressmaker now makes up a client's material. The customer weavers generally belonged to the cottager class, which is not well represented in the inventories.'

The fulling mills and corn mills of the medieval centuries demonstrated the potential of running water as a source of industrial energy; mills to the south of the region, at Cromford and Belper, offered the same message, and the extra efficiency provided by the overshot wheel was demonstrated to the Royal Society by John Smeaton in 1751. Sheffield had a centuries-old reputation as a centre of the cutlery industry; the main forges and workshops did not lie in the town itself, but in the peripheral valleys, where, according to long-established practices, water-wheels were harnessed to bellows at the furnaces and to the hammers of the forges. In the popular imagination, water-powered mills are associated with the first phases of the Industrial Revolution in the eighteenth century, and steam power with the steam-driven machinery of the nineteenth century. Although a few pioneering mills were established in the Calder valley, however, both water and steam made slower headway than might be imagined; domestic production increased around the marketing centres like Halifax, and, even in 1803, only one piece of cloth in 16 was woven in a mill. In 1793, a Mr Buckley announced his intention to build a steam-powered mill in Bradford, but was forced to withdraw after people in the affluent

neighbourhoods at the centre of the town campaigned against it on the grounds that it would produce unacceptable contamination by smoke and soot. Other innovations were more successful, with Cartwright's power loom being introduced in Doncaster in 1787 and then exported to the textile mills in Bradford. Taylor's shearing machines, which sheared the cloth after teazles had been used to raise the nap, were also introduced, though, like the power looms, they faced attacks from Luddites. In 1798, steam power was successfully introduced at Holme Mill in Bradford; there was no immediate conversion, but steam-powered looms were gradually established in factories in Yorkshire during the 1820s and, by 1835, Yorkshire had half of Britain's woollen mills and more than half the country's millworkers. Lawton and Pooley remark that: 'although the industrial revolution is often portrayed as an age of steam, steam-powered machinery and large-scale factory production were far from general at the end of its first phase ... The first Factory Inspectorate Report of 1838 showed that water still supplied 23 per cent of the power used in the cotton industry, despite the use of steam from 1784, and 43 per cent of the power used in woollen mills ...'. In 1800, the Bradford–Halifax–Keighley region had 22 water-powered worsted mills; by 1835, the region had 204 mills, most of them steam-powered, employing an average of 52 workers each. The main factor involved in this acceleration was probably a readiness to learn from the more advanced cotton producers in the neighbouring Manchester and Rochdale region; when the efficiency of steam power was improved, the factories in the West Riding were then able to exploit their superior access to coal reserves. Steam-powered worsted looms were introduced in 1824 and, by the middle of the century, Yorkshire had captured almost the entire worsted industry from Norfolk and lesser regions: Yorkshire had 87 per cent of worsted spindles and 95 per cent of worsted looms in 1850, and domestic looms had been put out of business. Perry notes that: 'more than their rivals Yorkshire entrepreneurs were prepared to innovate and adapt technology and organisation applied in the Lancashire cotton industry to Yorkshire wool – coal became important only at a later stage'.

The changes in manufacturing technology produced the most profound transformations in the pattern of settlement, which were evident at the town, village and countryside levels. Bradford was of scant significance during the eighteenth century, but it had a population of 44,000 by 1831, and in 1900 it had twenty times the population of a century earlier. Leeds was around the size of modern Harrogate in 1801, with a population of 53,000; in 1861 the population was 207,000 and, by the end of the century, it had a population of more than 400,000. The demand for industrial housing in the manufacturing towns of the West Riding resulted in the construction of hundreds of acres of congested and insanitary courts, terraced accommodation and back-to-backs, well before there was any legislation to protect social and public health interests. In the east of Leeds, housing built for workers towards the end of the eighteenth century was slotted into the medieval crofts running back from the roadside dwellings and was crammed into the yards of inns and the gardens of cottages; meanwhile, housing of a far higher quality was being built in the west. After 1780, back-to-back housing began to be built on adjacent green-field sites, sometimes with additional living space in cellars; such housing was built in Leeds for more than a century and a half, with the last of the notoriously badly ventilated back-to-backs being built as recently as 1937. In 1840, a committee investigating health in towns reported that, in Leeds, all the streets in one ward were: 'more or less deficient in sewerage, unpaved, full of holes, with deep channels formed by rain intersecting the roads, and annoying the passengers, sometimes rendered untenantable by the overflowing of the sewers and other more offensive drains . . .'. There was a general excess of industrial workers over accommodation, leading to the multiple occupation of space in lodging houses, with workers on one shift occupying beds vacated by those on another. In lodging houses in Leeds in 1851, there were an average of 2$\frac{1}{2}$ people per bed and an average of 4$\frac{1}{2}$ people to every room. Not all housing associated with the Industrial Revolution was of low quality, however; some of the cottages built for weavers during the eighteenth century were the best that had ever existed in their localities. Lawton and Pooley

write that: 'Some of the most active house construction in rural areas was aimed at industrial workers and not agricultural labourers. In the handloom textile and knitwear districts the long "weavers' windows" of often substantial stone or brick cottages testify to the impact of domestic industry. In addition the development of water-powered factory industry led to the growth of factory villages as manufacturers provided housing for workers attracted in from the surrounding countryside.'

Some industrial settlements in Yorkshire originated from the domestic industry. They began when near-destitute weavers set up home on the edge of the moor above an established village, existing as squatters and enclosing fields from the moor for their few beasts, while selling cloth at a nearby market or yarn to neighbouring weavers. The progress towards the establishment of a factory settlement could be a gradual one. The domestic industry was associated with vertical linkages, with spinners supplying yarn to the weavers, whose cloth was then passed on for dyeing and fulling, the same strands of fibre making several trips to and from a market like Halifax. (Some of the wool merchants were also manufacturers, buying and operating models of Hargreaves's 'spinning Jenny', which was adopted in Yorkshire in the 1770s and allowed one worker to spin several woollen threads at once.) Eventually, workers would recognise an interest in combining their work in a workshop or small factory, which might prosper until a capitalist arrived with the funds to establish a larger, more mechanised enterprise. This factory would be likely to catalyse the construction of one or more streets of terraced housing to accommodate the workforce. Such industrial housing could constitute a settlement in its own right, the best example – in all senses – being Saltaire, where more than 500 terraced dwellings were built beside the Aire by Sir Titus Salt in the 1850s and 1860s to provide superior housing and amenities for the employees of his textile mill. In other cases, the new industrial housing might be slotted into existing settlements, filling the gaps between the farmsteads and the cottages of the spinners and handloom weavers. Hey writes: 'The rows of two- or three-storeyed terraced houses that

characterized these new settlements were very different in scale and plan from the previous detached cottages.' Frequently, the changes would be profound. Holmfirth apparently began as a small settlement in the woods of Holm, and Hey records that: 'Until the late eighteenth century nearly all the inhabitants lived high above the river valleys; Holmfirth was a district name that originally denoted the hunting chase of the medieval lords of the huge manor of Wakefield. The valley bottom that is now occupied by the town of Holmfirth had few buildings apart from the manorial corn and fulling mills until 1784, when John Fallas, a woollen clothier acquired those properties and added a scribbling mill.' He quotes from Edward Baines's directory for 1822: 'The houses are scattered in the deep valley, and on the acclivities of the hills, without any regard to arrangement, or the formation of streets . . . The traveller, at his first view of this extraordinary village, is struck with astonishment at the singularity of its situation and appearance . . . This is a place of great trade, and the principal part of the inhabitants are employed in the manufacture of woollen cloth.'

Other textile industries were evident in the Dales, where flax and hemp were cultivated and cottage-based linen industries were developed. As noted above, the early phases of the Industrial Revolution were orientated to water power, and early nineteenth-century Nidderdale, where the Nidd provided a more reliable flow than the rivers of the limestone dales, supported a remarkable spectrum of industries: corn milling; the spinning of flax and cotton; the manufacture of linen and silk; the mining and smelting of lead; quarrying; and small-scale coal mining. At Glasshouses, a single flax mill employed some 264 workers in 1851. By this time, the textile industry had gravitated to water-powered mills, the latest stage in a transition from weaving, spinning or knitting as a cottage-based adjunct to farming, to one-loom farmers, then to two-loom smallholders, and then to the concentration of formerly independent operators in mills. The next stage witnessed the gradual migration of industry

Terraced industrial housing at Hebden Bridge

from the rural valleys as the adoption of steam-powered factory machines sucked manufacturing out of its rural backwaters and into the factory complexes on the coalfields.

Water was significant in terms of transport as well as of energy. The great era of canal building in Yorkshire spanned the first half of the Industrial Revolution and ran from 1750 to 1830. The relative importance of water transport during the medieval period is currently disputed, but York owed much of its earlier importance to its role as an inland seaport, and the Ouse system was navigated as far inland as places like Nun Monkton, on the Nidd. One pillar of Yorkshire's industrial success concerned its natural endowment with rivers like the Humber, Aire, Calder, Ouse and Don, which were navigable, or could be made more navigable, by building canals to bypass difficult sections. Under an Act of 1699, the Aire was made navigable to Leeds and the Calder navigable to Wakefield, so that the West Riding was connected to the North Sea, via the Humber, decades before the beginning of the Industrial Revolution. The region's waterways could then be linked to form networks, and, finally, the Yorkshire system could be linked to other networks by canals such as the ambitious Leeds and Liverpool canal, some 127 miles long, which was begun in 1770 and not completed until 1816, but which cut transport costs between Yorkshire and Lancashire by 80 per cent. With a population of 7,000 in 1801, Huddersfield's growth accelerated with its connection to the trans-Pennine canal system in 1811 by the Huddersfield canal, which crossed the mountains via a tunnel some 5,000 metres long under Standedge. The growth was enhanced by the introduction of steam power; by about 1835, Huddersfield and the neighbouring parish of Almondbury had 87 woollen mills, mostly making high-quality cloth.

Coal mining had a long history; coal was mined at Hipperholme in the thirteenth century – and had probably been mined for a few centuries before that at small pits in the West Riding. By the end of the medieval period, York was a coal- and peat-burning city, with coal being carted in on the roads from the west: had there been a better land transport system, then coal from the West Riding might have competed with the coal of the

North-East seaboard for a share of the London market. Medlicott notes that:

> The British coal industry underwent a period of rapid expansion after 1750, stimulated by the widespread adoption of coke in the iron smelting process and Henry Cort's reverberatory furnace. In particular, there was the development of the inland coalfields of the East and West Midlands, Lancashire and Yorkshire, following the construction of river navigations, canals and waggonways, where previously the prohibitive cost of moving coal had severely restricted sales and the exploitation of coal reserves . . . The period 1750 to 1830 was also one of technological improvements that allowed deeper-mined coal to be extracted and moved to the surface more efficiently. However, the coal industry was not transformed by a single technological innovation, but by a combination of improvements from several mine engineers from the major coalfields.

Great landowners had a substantial part to play in the dramatic development of coal mining in Britain during the Industrial Revolution. In the North-East, the Earl of Durham was heavily involved in mining; in the West Midlands, Lord Dudley developed a great mining and industrial empire, and the Marquis of Bute invested heavily in coal-handling facilities at Cardiff docks. In Yorkshire, the Dukes of Norfolk had 20,000 acres of land in the vicinity of Sheffield, sharing a boundary with the 15,000-acre estate of the Rockingham-Fitzwilliam family, and three seams, containing a rich variety of steam, coking, gas and domestic coal. Close to the coal reserves lay beds of ironstone, so that the various essentials for industrial development were concentrated in the locality: 'The collieries provided fuel for the furnaces and placed regular orders for iron goods. The level of capital investment required for large-scale mining became increasingly beyond the means of the sole capitalist entrepreneur of the mid-eighteenth century, and it was at this point that the great land-owner took on the role of colliery proprietor.' The first railway in Yorkshire was built in 1834, linking Leeds to Selby (and, subsequently, to Hull). The arrival of the steam engine, whether harnessed to haul a train of trucks or carriages, or to power a factory, marked a new phase in the Industrial Revolution. The changes that resulted would rebound around Yorkshire, producing all manner of consequences. The emergence of a relatively mobile army of industrial workers gave birth to the institution of the family seaside holiday, underwriting, for a century or so, the fortunes of Scarborough, Bridlington, Filey and Saltburn. In the coalfield towns, the rate of urban growth would be unprecedented, but it would be achieved at the cost of a bleeding of population from the water-powered mill sites and the cottage bases of domestic industry lying in the Dales, far from the coalfields. In due course, the countrymen and their offspring who had survived enclosure by obtaining work as farm labourers would be put out of work by the machines produced in the new factories and workshops; the empty places in the countrysides would then be filled by commuters who were several generations removed from the land and total strangers to its culture. Today, a society which has scarcely come to terms with the shock of the urban/industrial explosion faces the uncharted mysteries of the post-Fordist world.

REFERENCES

S. D. Chapman (ed.), *The History of Working Class Housing* (Exeter: 1971).

M. W. Flynn, *The History of the British Coal Industry*, vol. 2 (Oxford: 1984).

D. Fraser (ed.), *A History of Modern Leeds* (Manchester: 1980).

M. Freeman, 'The Industrial Revolution and the regional geography of Britain: a comment', *Transactions of the Institute of British Geographers*, NS 9 (1984), pp. 507–12.

D. Gregory, 'The process of industrial change 1730–1900', in R. A. Dodgshon and R. A. Butlin (eds), *An Historical Geography of England and Wales* (London: 1978), pp. 291–311.

D. Gregory, *Regional Transformation and Industrial Revolution* (London: 1981).

D. Hey, *The Making of South Yorkshire* (Ashbourne: 1979).

D. Hey, *Yorkshire from AD 1000* (Harlow: 1986).

P. Hudson, *The Genisis of Industrial Capital of the West Riding Textile Industry 1750–1850* (Cambridge: 1989).

P. Jackson, 'The racialisation of labour in post-war Bradford', *Journal of Historical Geography*, 18 (1992), pp. 190–209.

D. T. Jenkins and K. G. Ponting, *The British Wool Textile Industry 1770–1914* (London: 1982).

B. Jennings (ed.), *A History of Nidderdale* (Huddersfield: 1967).

R. Lawton and C. G. Pooley, *Britain 1740–1950: An Historical Geography* (London: 1992).

I. R. Medlicott, 'The development of coal mining on the Norfolk and Rockingham-Fitzwilliam estates in South Yorkshire 1750–1830', *The Yorkshire Archaeological Journal*, 59 (1987), pp. 103–18.

P. J. Perry, *A Geography of 19th Century Britain* (London: 1975).

G. N. von Tunzelmann, *Steam Power and British Industrialisation to 1860* (Oxford: 1982).

Index

Location references in *italics* denote an illustration.